Naturalists' Handbooks 14

Mosquitoes

KEITH R. SNOW
School of Science,
Polytechnic of East London

With illustrations by the author
and coloured plates by A. J. E. Terzi

T0306740

Pelagic Publishing
www.pelagicpublishing.com

Published by Pelagic Publishing
www.pelagicpublishing.com
PO Box 725, Exeter, EX1 9QU, UK

Mosquitoes
Naturalists' Handbooks 14

Series editors
S. A. Corbet and R. H. L. Disney

ISBN 978-1-78427-047-6 Paperback
ISBN 978-1-78427-048-3 Hardback

Digital reprint edition of:
ISBN 0-85546-275-2 (1990) Paperback
ISBN 0-85546-276-0 (1990) Hardback

Colour plates by A. J. E. Terzi reproduced
by courtesy of The Natural History Museum

British Library Cataloguing in Publication Data
A catalogue record for this book is available from the British Library.

Contents

Editors' preface

Students at home or university, and others without a university training in biology, may have the opportunity and inclination to study local natural history but lack the knowledge to do so in a confident and productive way. The books in this series offer them the information and ideas needed to plan an investigation, and practical guidance to help them carry it out. They draw attention to regions on the frontiers of current knowledge where amateur studies have much to offer. We hope readers will derive as much satisfaction from their biological explorations as we have done.

Because mosquitoes carry disease, tropical forms have received much attention from entomologists. Few ecologists have studied mosquitoes in Britain, perhaps partly because identification has been difficult. The key published in 1987 by the Freshwater Biological Association will make British mosquitoes much more accessible as subjects for ecological research. We hope this book will also help to make them attractive to a wider audience. Effective control of mosquitoes as vectors of disease depends on a sound understanding of their natural history. Studies on British species can contribute to that understanding.

S.A.C.
R.H.L.D.

Acknowledgements

My grateful thanks are due to Clem Ramsdale, Dr Alun Rees, Dr Mike Service and Bruce Townsend for constructive comments on sections of the text and to John Lane for access to the mosquito collections of the London School of Hygiene and Tropical Medicine.

I also wish to express my thanks to the series editors Dr Sally Corbet and Dr Henry Disney for their advice and assistance in the preparation of this book

I am also extremely grateful for permission to reprint figures 2, 4, 5, 10–12, 16, 19, 33–35 and 38 from the Freshwater Biological Association *Keys to British Mosquitoes* and figures 15, 17 and 25 from *Ecological Entomology, Entomologist's Gazette* and *Entomologist's Monthly Magazine* respectively. Figures 22–24 and 28–32 were based upon illustrations in *Manual on Practical Entomology in Malaria* (1975) *WHO Offset Publication No. 13* and were drawn with permission from the World Health Organization. Finally my thanks to the Natural History Museum for allowing me to reproduce the colour illustrations by A.J.E. Terzi.

K.R.S.

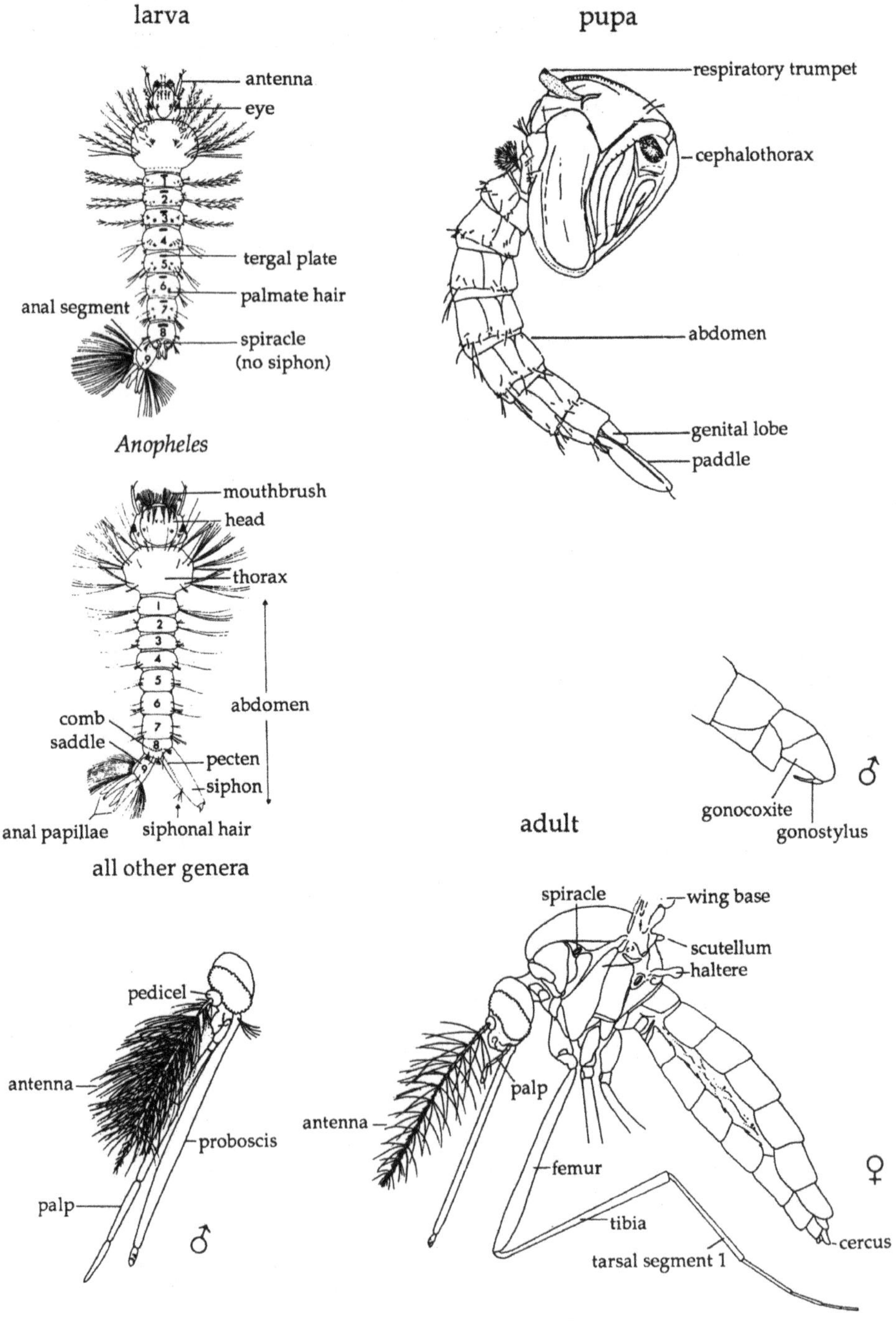

Fig. 1. The illustrations above show many of the features used in mosquito identification. Although they appear in the text, they are grouped together here for ease of reference.

1 Introduction

There is a tendency to think of mosquitoes as tropical insects even though they are found throughout the world and are abundant in many parts of Britain. Mosquitoes are often colloquially referred to as 'gnats'. We are all familiar with their 'bites', and many people can recognise them in both their aquatic and terrestrial life phases. However, it is essential to be aware that there are several closely related groups of insects with which they can be confused, especially in the aquatic stages.

Because of the immense importance of mosquitoes as vectors of parasitic infections, such as malaria, many tropical species have been studied in great detail. By contrast, our knowledge of the British species is very patchy. Indeed there is great scope for the amateur and student to fill the gaps in our knowledge, as well as for the professional entomologist to plan and conduct more elaborate field and laboratory experiments. With the study of mosquitoes in Britain there is something for everyone and all contributions count.

Classification of mosquitoes

Mosquitoes belong to the order Diptera, the two-winged flies. They are members of a single family, the Culicidae, which is divided into three sub-families: Anophelinae, Culicinae and Toxorhynchitinae. In Britain there are five species of Anophelinae, all in the genus *Anopheles*, and 27 species of Culicinae in five genera: *Aedes*, *Coquillettidia*, *Culex*, *Culiseta* and *Orthopodomyia* (p. 2). There are no representatives of the Toxorhynchitinae in Britain.

Of the 32 recorded species, five must be considered doubtfully native as they have not been found for many years: *Aedes communis*, *Aedes leucomelas*, *Aedes sticticus*, *Culex modestus*, and *Culiseta longiareolata*. However, with the recent renewed interest in British mosquitoes, and the establishment of the British Mosquito Recording Scheme in 1980, these rare mosquitoes may be rediscovered and other species, as yet unrecorded, may be added to the British list. For example, *Aedes vexans* has been reported once more (Ramsdale & Smith, 1990) and *Anopheles algeriensis*, previously known in Britain only from Norfolk in the 1930s–1950s, was recently discovered in Anglesey trying to bite a member of the recording scheme. Subsequently the larval habitats were found (Rees & Rees, 1989). *An. algeriensis* resembles *An. claviger* closely and may exist, overlooked, in other areas.

Life cycle of mosquitoes

Culex pipiens, probably the most widespread and abundant mosquito in Britain, will serve to demonstrate the life cycle of a mosquito. Other species showing variations from this pattern are discussed later in this chapter.

The classification of British mosquitoes.

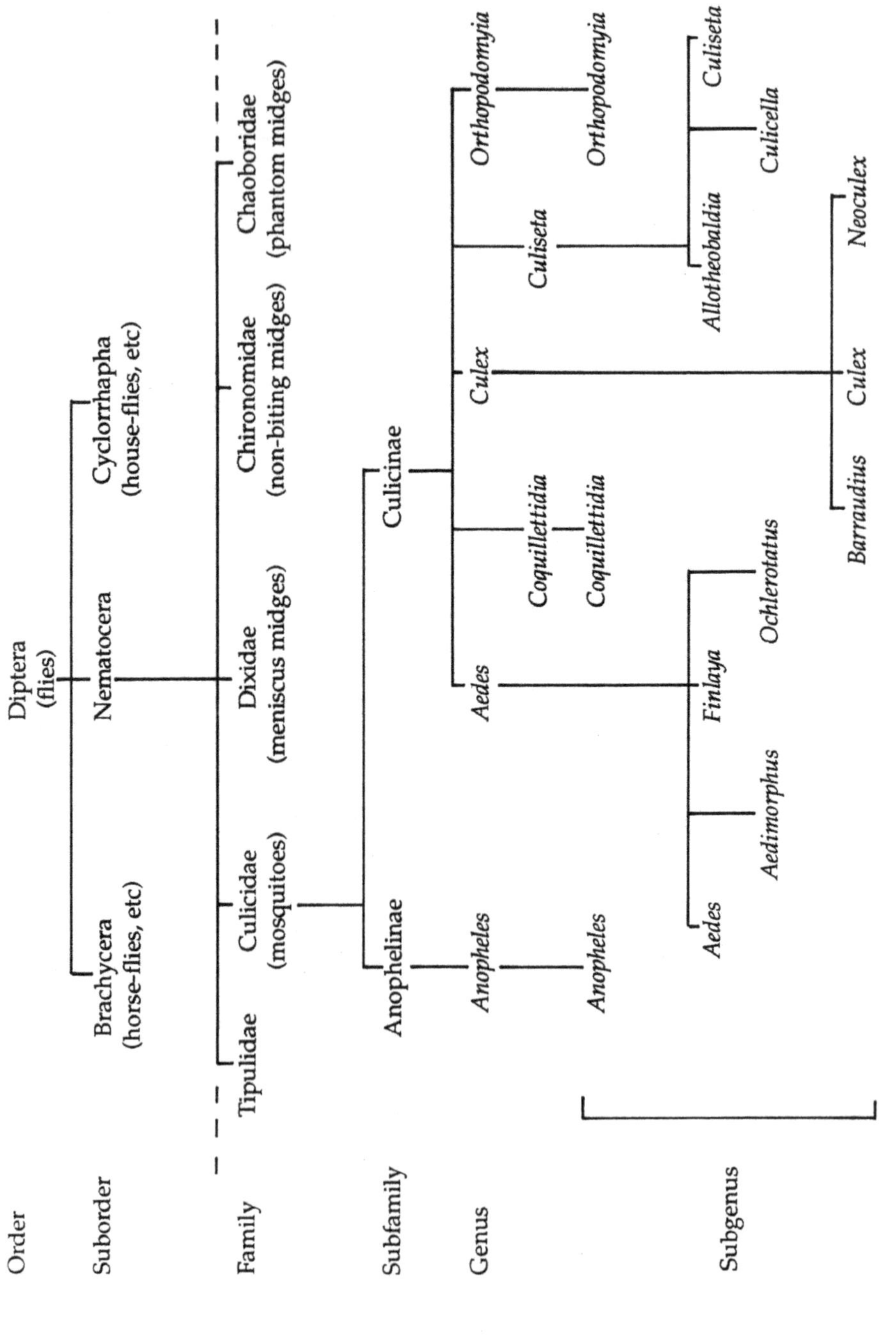

Eggs of *Cx. pipiens* are laid in batches forming rafts on the surface of still water from early April onward. A tiny larva hatches from the egg and feeds by creating a current with its mouthbrushes. It breathes by taking in air through a tube, the siphon, at the rear of its body. Some days later the larva moults to become the next larval stage and this is repeated until the fourth larval stage is reached. By mid-May this final larval stage moults to become a pupa. This also breathes air but does not feed. Within a week or so an adult emerges from the pupal skin.

Mating takes place soon after emergence and both males and females begin to feed on nectar and other plant juices which provide them with sugars for daily energy needs. The females also seek a blood meal, usually from birds in the case of *Cx. pipiens*. This provides them with the nutrients necessary for egg development.

From August onwards emerging females mate but do not blood-feed. Instead they feed on plant juices to build up an energy store in the form of a fat body. This reserve will allow them to survive the winter in a cool place such as an outbuilding or attic. No males survive the winter but as spring approaches the fertilised females become active once more, seek a blood meal and lay their eggs. Overwintering females are rarely seen later than May.

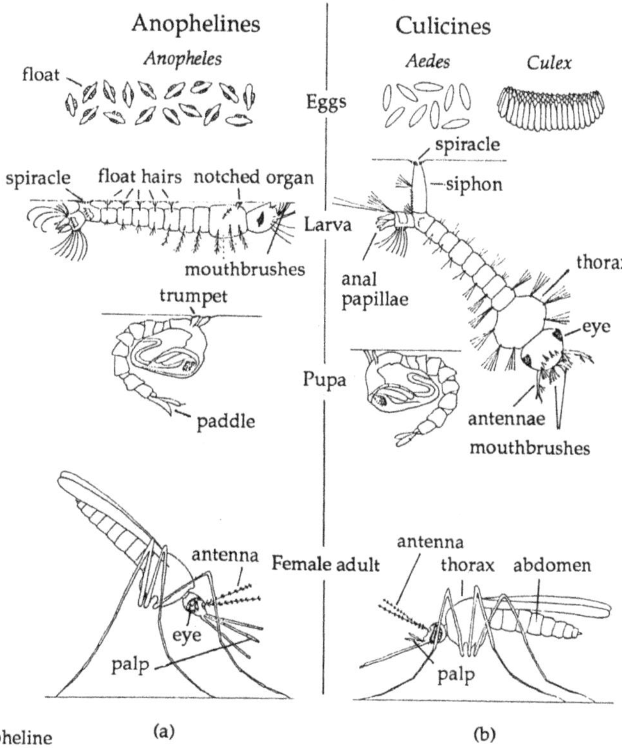

Fig. 2. Life stages of (a) anopheline and (b) culicine mosquitoes.

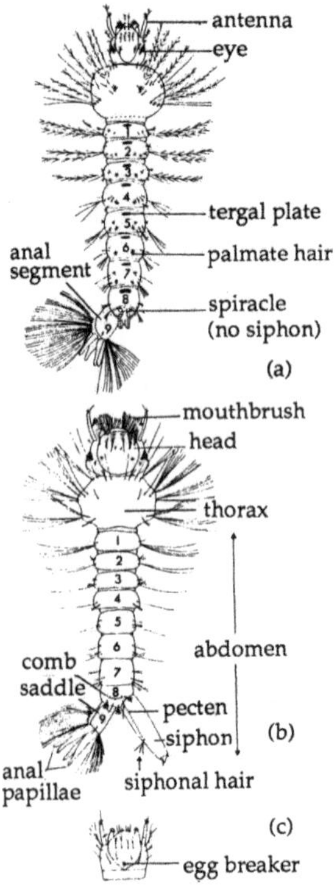

Fig. 3. (a) Anopheline larva; (b) culicine larva; (c) head of first-instar larva showing egg breaker (all viewed from above). In (a) and (b) the anal segment (abdominal segment 9) has been displaced to show its features more clearly.

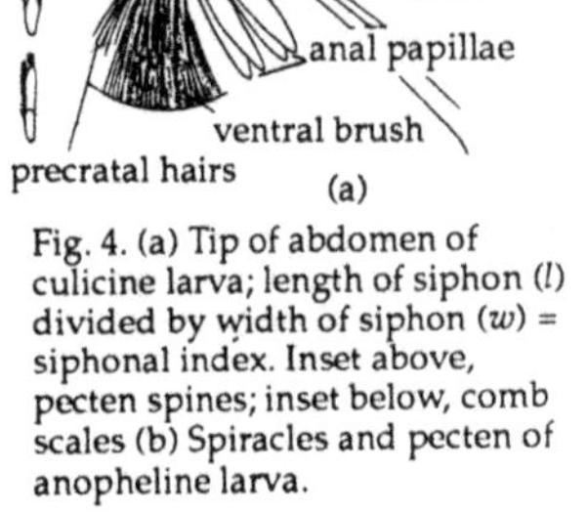

Fig. 4. (a) Tip of abdomen of culicine larva; length of siphon (*l*) divided by width of siphon (*w*) = siphonal index. Inset above, pecten spines; inset below, comb scales (b) Spiracles and pecten of anopheline larva.

Where to find mosquito eggs, larvae and pupae

The aquatic stages of mosquitoes may be found in a variety of habitats. Some species live in permanent ground water while others are found in temporary woodland pools and water-filled ditches, flooded meadows and freshwater marshes. Yet others develop exclusively in water-filled tree-holes or brackish and saline waters. Some species are catholic in their choice of sites and may select almost any aquatic site including artificial accumulations of water in rain barrels, animal troughs, buckets and cans.

How to recognise a mosquito at all stages of its life cycle

Eggs of mosquitoes are generally elongate-oval, brown or blackish and up to 1 mm in length. Whether laid separately or in compact rafts they vary considerably in form: those of *Anopheles*, for example, are boat-shaped and have a pair of air-floats at the sides (fig. 2); those of *Orthopodomyia* have a narrow frill; *Culex* eggs have a prominence called a corolla while those of *Aedes* are without surface projections (fig. 2). When magnified, mosquito eggs are seen to have patterned surfaces which may often be used in their identification.

The first-instar larva, which hatches from the egg, is at once distinguished from subsequent larval instars, not only by its small size, but also by the possession of an egg breaker on the top of the head (fig. 3). All four instars are similar, varying in size and in the number, arrangement and complexity of certain hairs and other surface structures, for example the form and number of pecten spines (table 1).

Mosquito larvae (fig. 3) do not possess legs and their bodies are divided into a well developed head bearing mouthbrushes and antennae, a bulbous thorax which is broader than the head and abdomen, and an elongate abdomen which ends in an anal segment bearing four gill-like lobes, the anal papillae. The next-to-last abdominal segment bears either a pair of openings to the respiratory system, the spiracles (*Anopheles*) (fig. 4b), or an elongate respiratory siphon (remaining genera) (fig. 4a).

Larvae of another dipteran family, the Chaoboridae or 'phantom midges', are easily confused with mosquito larvae. Chaoborid larvae are recognised by their long antennae with which they capture aquatic prey. In contrast, the antennae of mosquito larvae are short and are not used in feeding. Chaoborid larvae also possess conspicuous pigmented air sacs, the hydrostatic organs, one pair in the thorax and another pair in the seventh abdominal segment (fig. 5). These organs appear silvery because of the air in them and allow chaoborid larvae to control their buoyancy and to swim at various depths in a horizontal position. Mosquito larvae do not possess hydrostatic organs and generally respire at the water surface, submerging to escape danger and, in some species, to feed by browsing. At the water surface they lie horizontally (*Anopheles*) or are

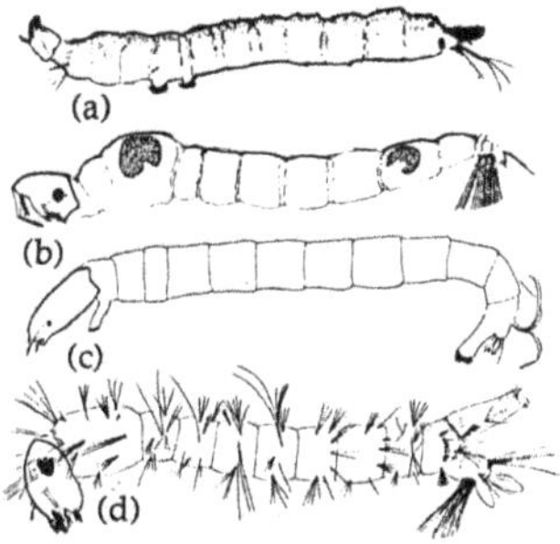

Fig. 5. Larvae of (a) Dixidae, (b) Chaoboridae, (c) Chironomidae and (d) Culicidae (Culicinae).

Table 1. *The number of pecten spines and comb scales in the larval instars of* Aedes punctor

Instar	Number of pecten spines	Number of comb scales
1	5–8	5–8
2	10–15	10–16
3	13–16	12–17
4	11–23	8–29

suspended more or less vertically by the siphon (culicines) (fig. 2, p.3). *Coquillettidia richiardii* is an exception as the larvae attach to the roots and stems of aquatic plants by modified respiratory siphons (fig. 6a), using oxygen from the air spaces of the plants.

The pupa of a mosquito is recognised by its distinctive comma shape, the 'dot' of the comma representing the combined head and thorax (cephalothorax) and the 'tail' the abdomen (fig. 7). The pupae remain at the water surface for most of the time, breathing through their paired respiratory trumpets (fig. 7). When disturbed they dive with a jerking motion, returning to the surface passively. *Coquillettidia richiardii* is once again an exception; its pupa attaches to submerged parts of plants by means of modified trumpets (fig. 8) with which it obtains air from the air spaces of the plant. The sex of a pupa can be determined by examination of the genital lobe which contains the developing adult external genitalia. In the female this is short and undivided whereas in the male it is long and consists of two lobes (fig. 9). Next to the genital lobe are the paddles (fig. 9) which are flattened structures, stiffened by a single central midrib.

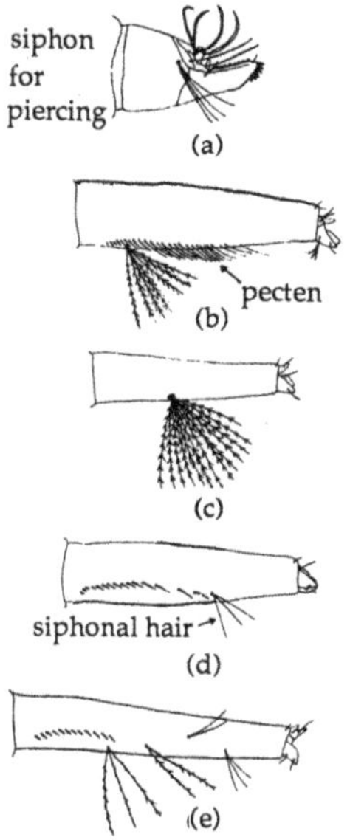

Fig. 6. Side view of the siphons of (a) *Coquillettidia*, (b) *Culiseta*, (c) *Orthopodomyia*, (d) *Aedes* and (e) *Culex*.

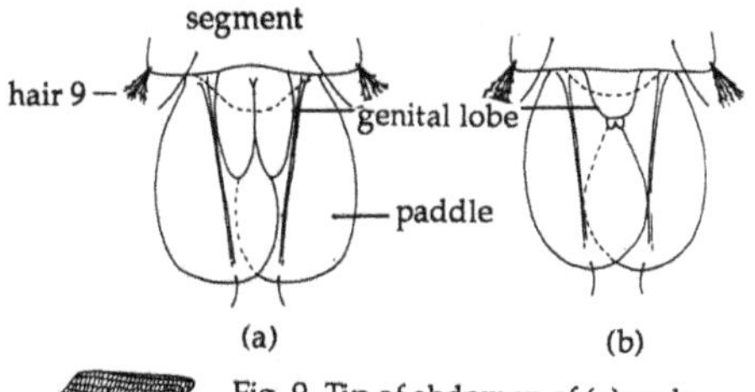

Fig. 9. Tip of abdomen of (a) male and (b) female mosquito pupa from below showing the shape of the genital lobe.

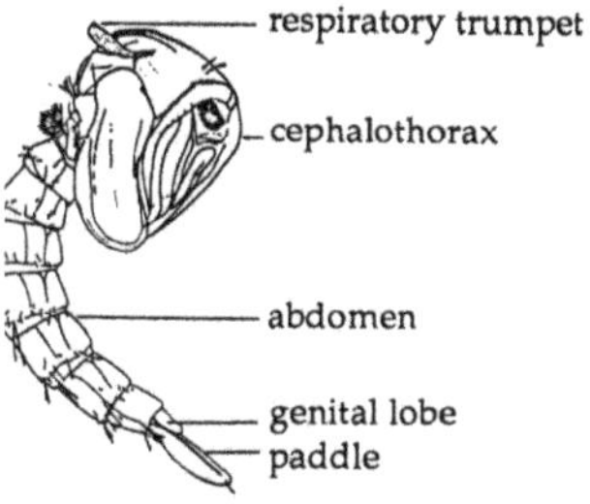

Fig. 7. Side view of mosquito pupa.

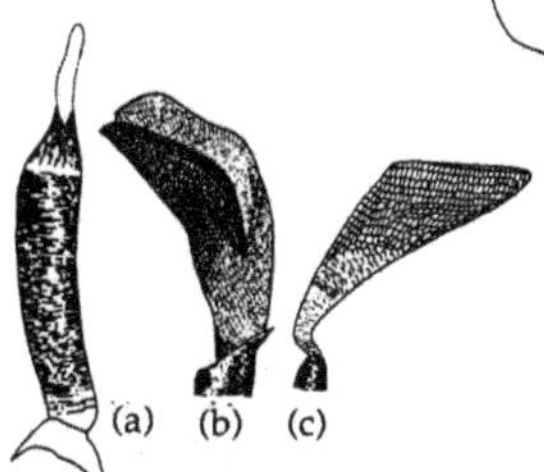

Fig. 8. Respiratory trumpets of (a) *Coquillettidia*, (b) *Anopheles* and (c) *Orthopodomyia*. The tip of the trumpet in *Coquillettidia* (dotted line) is often broken off.

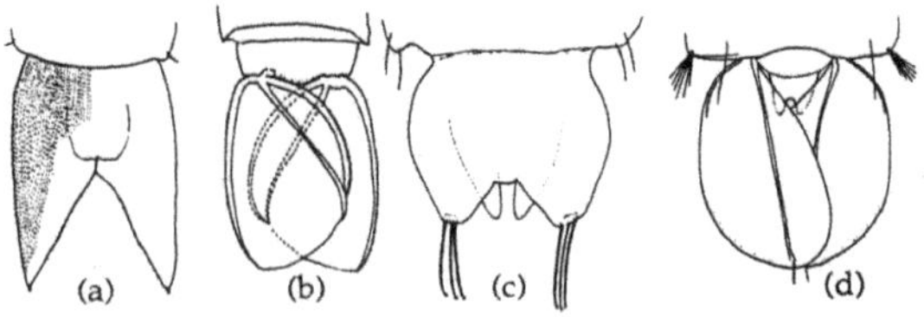

Fig. 10. Tip of abdomen of pupa of (a) Dixidae, (b) Chaoboridae, (c) Chironomidae and (d) Culicidae.

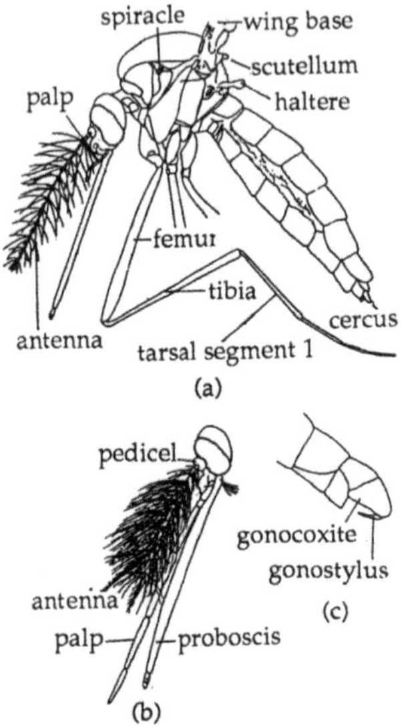

Fig. 11. Characters of adult mosquitoes. (a) female; (b) head of male; (c) tip of male abdomen.

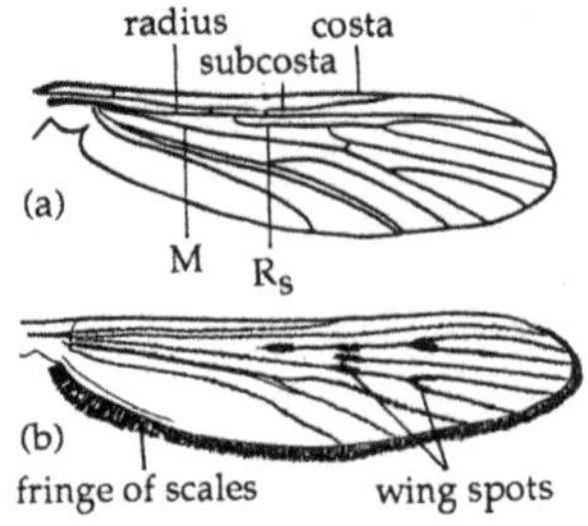

Fig. 12. Mosquito wing (a) showing veins and (b) showing spots (*Anopheles atroparvus*).

The pupae of the families Chaoboridae (phantom midges), Dixidae (meniscus midges) and Chironomidae (non-biting midges) are rather like those of mosquitoes. However, they can easily be distinguished by examining the end of the abdomen. The pupae of mosquitoes, chaoborids and chironomids all have paddles (fig. 10) but these are supported by three ribs in chaoborids, one in mosquitoes and none in chironomids. The abdomen of dixids ends in tapered lobes (fig. 10)

Adult mosquitoes may immediately be distinguished from other flies by their long proboscis (fig. 11), their wing venation (fig. 12) and the presence of scales on the wings, legs, thorax and other parts of the body.

Mosquitoes are slender and relatively small insects, measuring between 4 and 10 mm in length. The head bears a pair of large compound eyes (but no simple eyes) and a pair of long, threadlike antennae. Each antenna consists of a large, globular basal region and a filament of 13 segments. In females each segment bears whorls of short hairs (the antenna is pilose) while in males the hairs are long giving the male antenna a bushy (or plumose) appearance (fig. 13). Close to the antennae arise the palps which vary in shape depending on the sex and subgroup of the mosquito (fig. 13).

Importance of mosquitoes in Britain

Whilst feeding, female mosquitoes secrete saliva into our skin and the chemicals within this evoke a strong local skin reaction in many people. These 'bites' occur most frequently in May to September, although a few species of mosquito overwinter in houses and may feed throughout the year.

In the past, malaria occurred naturally in Britain, being transmitted in the valleys and marshlands where suitable mosquito vectors were found. The most efficient vector was *Anopheles atroparvus* and this was the chief source of the benign tertian form of malaria occurring in these islands. The disease was commonly called 'ague' and was especially rife in the marshlands and estuaries of East Anglia, Essex, Kent and counties of the south coast. The drainage of the marshlands, housing improvements, climatic changes and availability of the anti-malarial drug, quinine, led to the decline of ague in Britain, and by the beginning of this century it had been eliminated.

During the First World War (1914–18), transmission of malaria occurred once again as thousands of soldiers returned to Britain with malaria. By 1921 the number of reports of transmission had declined to only a few cases per year. There was a second influx of imported cases during and after the Second World War (1939–45).

Nowadays about 2000 cases of imported malaria are detected each year, all contracted abroad. However, during exceptionally hot summers malaria may be contracted locally. In 1983, for example, two cases of malaria appeared near Gatwick Airport in people who had not travelled to malarious areas and were, almost certainly, bitten by infective tropical *Anopheles* mosquitoes which had entered this country aboard aircraft.

Another mosquito-borne disease, yellow fever, broke out on ships arriving at British ports in the last century. It would appear that some of these ships also imported the tropical mosquito, *Aedes aegypti*, probably breeding in water tanks on the decks. After feeding on sailors infected with yellow fever virus the mosquitoes subsequently bit people in the port towns, so transmitting the virus. It is unlikely that these conditions could be duplicated now and hence it is doubtful whether yellow fever could occur in Britain again.

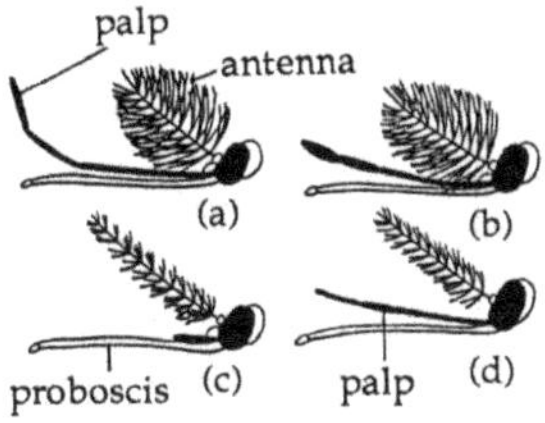

Fig. 13. Mosquito heads. (a) Male culicine; (b) male anopheline; (c) female culicine; (d) female anopheline.

Many other virus diseases are transmitted by mosquitoes in Europe. An example is Inkoo virus which is widespread in Finland, where it is transmitted by species also present in this country such as *Aedes punctor*. Most cases of this disease are without symptoms but some patients report mild fevers. It is of interest that the virus causing myxomatosis in rabbits may also be carried by mosquitoes, although its normal vectors in Britain appear to be fleas.

Control of mosquitoes

If mosquitoes become an intolerable nuisance then a number of control methods are available. These range from the removal or modification of the offending larval habitats to make them unsuitable for mosquitoes, through the use of fish predators to the use of chemical and microbial insecticides. In the tropics an additional control method involves the release of specially bred sterile male insects with the aim of achieving sterile matings and so reducing the wild population. Such a method is expensive and not warranted in Britain, where mosquitoes do not pose a serious risk to health.

A recent survey of mosquito nuisance and control in this country (Snow, 1986, 1987c) showed that most Local Authorities conducted their control efforts against the aquatic stages. The methods used included land drainage, infilling, the removal of water containers, flooding areas of marshlands with sea water and the application of pesticides to larval habitats. Adults are controlled by spraying residual insecticides on walls inside buildings.

A wide range of chemicals has been used in the past against mosquito larvae, among them DDT, lindane, malathion and oil. Nowadays no conventional insecticide is approved for application to water for the control of mosquitoes in Britain. However, formulations of the mosquito-attacking bacterium *Bacillus thuringiensis* var. *israelensis* (Bti) and the newly developed surface active materials, the lecithins, are permitted. A comprehensive list of approved pesticide products and chemical compounds is now contained in an H.M.S.O. publication (Reference Book 500) published annually.

Commercially produced lecithins are self-spreading liquids which, when applied to water surfaces, produce a non-toxic monomolecular layer (monolayer) which reduces the surface tension of the water, wetting the larval siphons and respiratory trumpets of the pupae which then cannot adhere to the surface film and so drown (see Guthrie, 1989). It also prevents the successful emergence of adults. Although lecithins kill pupae very rapidly they may require several days to kill larvae, especially early instars which can respire through their cuticle while submerged. The frequency of applications depends on how long the monolayer remains intact without being removed by rainfall, physical disturbance or biodegradation, and the application rate varies with the degree of pollution of the water.

Products based upon the bacterium *Bacillus thuringiensis* var. *israelensis* have been shown to be both effective and environmentally safe. The spores of this bacterium produce protein crystals which, when eaten by mosquito larvae, become activated to produce toxins which attack the gut wall causing death. Bti affects only a very narrow range of insects and acts quickly, normally within several hours. Because it needs to be eaten it has no effect on pupae.

It is possible to combine the lecithins and Bti to produce a very effective insecticide, the monolayer spreading the Bti evenly over the surface of the water. The two products are considered to be complementary, Bti being most active against early-instar larvae and the monolayer having its major effect on older larvae and pupae.

Personal protection

Those who study adult mosquitoes face a dilemma. It is necessary to attract the mosquitoes for collection, and yet nobody wants to be bitten. This problem may be partially overcome by wearing suitable clothing so that only the hands and face remain exposed. However, it must be realised that mosquitoes can bite through thin clothing, especially when it is tight-fitting and in close contact with the skin.

At other times the priority is to keep mosquitoes at bay. Here chemical insect repellents are of value. Repellents suitable for application to the skin and effective against mosquitoes include dimethylphthalate, diethyltoluamide ('Deet') and ethylhexadiol. Citronella oil is also used and has a very pleasant smell. Although it appears to be effective when freshly applied to the skin it evaporates rapidly and so gives only short-term protection. For maximum protection repellents must be applied frequently and all exposed parts of the body (excluding eyes and lips) must be treated. Even so, repellents only reduce the chances of being bitten and do not afford long-term protection. Schreck (1985) found that pure 'Deet' gave protection for 2 hours against *Anopheles albimanus* but that lower concentrations (25–75%) provided little more than 30 minutes' protection. Most commercially available repellents fall into this range of dilution, a few exceeding it slightly. Repellents may be purchased in the form of lotions, creams, gels, wipes or solid sticks.

A word of warning: some people, especially young children, may experience allergy and other adverse reactions to repellents and many repellents soften plastics (such as spectacle frames), certain synthetic fibres and rubber.

Insect repellents may be used to impregnate clothing to give additional protection against biting insects. Such clothing should be kept in air-tight containers when not in use and be re-impregnated after laundering.

The search for newer and better chemical repellents continues but is hampered by the fact that we do not know how they work. It has been suggested that they may interfere with the perception of host-attractant signals, by exciting a receptor for a competing behaviour, by switching a sensory message from attraction to repulsion, by activating several different receptor systems so that the repellent, in effect, 'jams' meaningful sensory information, or by exciting a repellent (noxious substance) receptor (Davis, 1985).

On occasions, mosquitoes may become a nuisance in houses in this country. If so, aerosol insecticide sprays are effective together with thermal vapourisers which incorporate a small tablet of a synthetic pyrethroid (allethrin) which is slowly released into the air. Smouldering mosquito coils, which contain pyrethrum, are also effective. Electronic buzzers which emit a high-pitched sound have been claimed to repel mosquitoes but they have been shown to be ineffective (Foster & Lutes, 1985). Two brands of buzzer sold as mosquito repellents have now been taken off the British market after prosecutions under the Trades Descriptions Act (Curtis, 1986). In fact, unlike male mosquitoes, females show little or no response to sound.

2 Life histories

During its life cycle a mosquito passes through the following stages: egg; first-, second-, third- and fourth-instar larvae; pupa; adult (fig. 14).

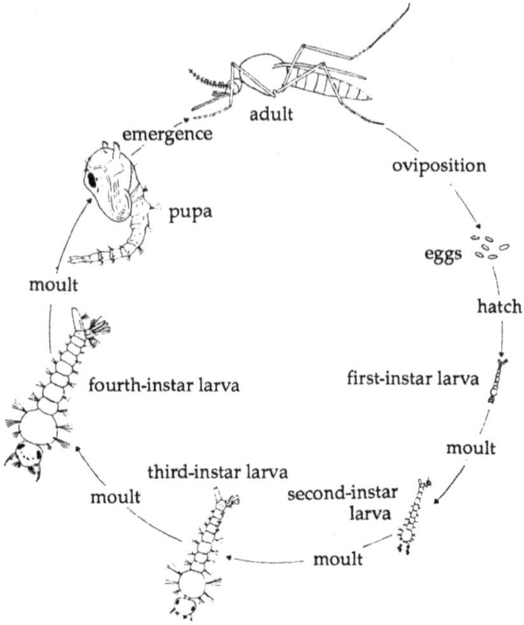

Fig. 14. Life cycle of a mosquito.

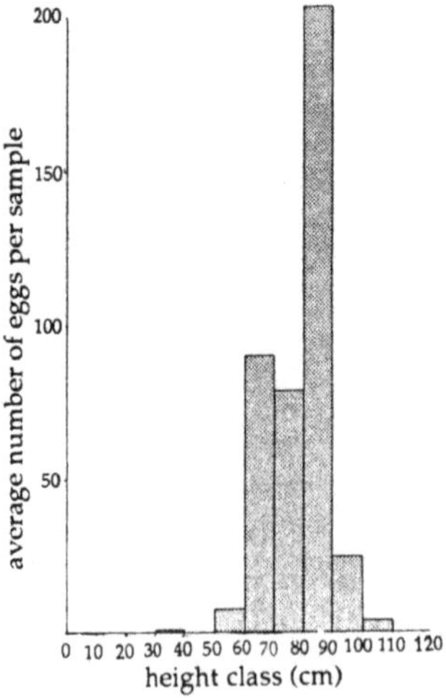

Fig. 15. Histogram showing the average number of eggs of *Aedes punctor* per standard sample for each height category, measured from the rim of a dry hollow 122 cm deep.

Eggs

Depending on the species, 30–350 eggs are normally laid in each egg batch. Marshall (1938) quotes the largest recorded number of eggs in a batch as 433 for *Culex pipiens* and notes that a *Culiseta subochrea* female laid five egg rafts within a period of 33 days, producing a total of 1036 eggs, only six of which failed to hatch.

The eggs are laid either singly (*Aedes*, *Anopheles*, *Orthopodomyia* and *Culiseta* subgenus *Culicella*) or in batches of several hundred, forming rafts (*Culex*, *Coquillettidia* and *Culiseta* subgenus *Culiseta*). In most cases the eggs are laid on the surface of still or slowly moving water but females of the genus *Aedes* and the subgenus *Culicella* deposit their eggs either in dried-up hollows and ditches subject to winter flooding, or above the water line (fig. 15) in places which will later be submerged (Fallis & Snow, 1983b). As the water level rises in these areas the eggs become immersed and eventually hatch. In years when rainfall is low not all of the eggs may be submerged and some may remain viable and hatch the following year.

Although laid in rafts, eggs of *Culex territans* are deposited above the water and on hatching many of the eggs detach and drop into the water. Alternatively the young larvae wriggle down from their egg-shells to enter the water.

oviposition: egg-laying

In those mosquito species which oviposit in permanent waters the eggs hatch as soon as the embryo is sufficiently developed. Eggs of other species, such as *Aedes*, require conditioning by a sequence of stimuli before hatching will occur. This entails exposure to freezing or near-freezing conditions followed by immersion in water. The larvae hatch in response to a slow decrease in the oxygen content of the medium caused by microbial activity (Fallis & Snow, 1983a).

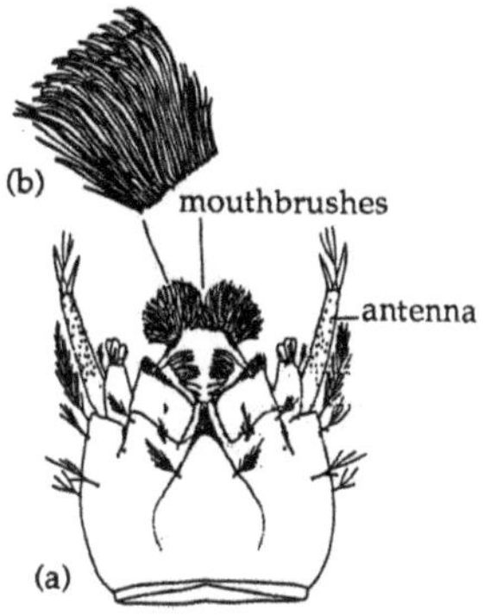

Fig. 16. (a) Head of larva, viewed from below (b) mouthbrushes enlarged.

Larvae and pupae

The newly hatched first-stage larva escapes from the egg aided by an egg-breaker on the top of the head (fig. 3, p. 4). Larvae feed by producing water currents with their moustache-like mouthbrushes (fig. 16) and sieving food from the water. All *Anopheles* larvae collect their food at the water surface but other species hang down from the surface to feed. In addition to filter-feeding, many species browse on leaves, algae and bacteria deeper in the water, remaining submerged for several minutes. A very fruitful line of investigation would be to see whether diet and feeding behaviour vary with age. Diet may be investigated by direct observation, examination of gut contents (using a compound microscope) and experiments with different combinations of food items. Some larvae are known to feed on minute particles and organisms filtered from the water. Older larvae, in particular, have been reported browsing filamentous algae and dead leaves.

exuviae: the skin or cuticle shed by an insect during moulting. Exuviae is a plural noun and so remains unchanged whether used in the singular or plural

instar: the stage in the development of an insect between two moults. It is a first-instar larva that hatches from the egg

In all there are four larval stages (fig. 14) and each moults to produce the next instar, casting its skin (exuviae) which retains all of its surface characteristics. The final larval stage (fourth instar) moults to become an active, comma-shaped pupa which does not feed.

The time spent in the larval stages is variable and depends on the species and external influences such as food supply and water temperature. Depending on conditions, the total time for development from hatch to pupation ranges from about 10 days to at least 6 months, being extended further in those species which overwinter as larvae.

Mosquito larvae and pupae are air-breathers and are found at the water surface for much of the time. An exception to this is *Coquillettidia richiardii* which as larva and pupa remains submerged, attached to the roots and stems of certain aquatic plants to obtain oxygen from them. It is of interest that, under laboratory conditions, both larvae and pupae of the tropical mosquito, *Aedes aegypti*, have been observed with their breathing tubes inserted into submerged air bubbles. When water weeds are placed in glass beakers the oxygen produced often forms large bubbles which attach to the imperfections in the glass. Larvae and pupae use the bubbles as 'diving bells' and may remain submerged for up to 30 minutes. To what extent this happens in the wild is not known. The pupal stage lasts for 4 or more days and then the skin (exuviae) splits along the back and the adult mosquito emerges. Having freed itself from the pupal skin the adult rests for a short while on the water surface, the discarded pupal skin or adjacent vegetation. This allows time for the wings to unfold and harden sufficiently for flight.

Mating

Mating usually occurs close to the larval habitat often shortly after emergence. Observations have shown that males can mate as soon as the rotation of their external genitalia is complete, about 12–24 hours after emergence. In some species swarms of males form, females entering the swarms for pairing. However, in other species swarms have not been observed and the *molestus* form of *Culex pipiens* is known to be able to mate in confined spaces. Presumably these do not form swarms.

In *Aedes cantans*, for example, male swarms consist of up to about 100 individuals, although typically less than half this number. They congregate a metre or so above the ground in the early evening and disperse before sunset (fig. 17) in late May to early August (Fallis & Snow, 1984). Additional information on the swarming behaviour of British *Aedes* mosquitoes is given by Gomulski (1988).

Swarms of *Cq. richiardii* have been observed from mid-June until the end of August, always close to larval habitats. They form after sunset and continue into the dark and then reform immediately after sunrise for a short time (Marshall, 1938).

At mating, the female receives enough sperm to fertilise most or all of the eggs laid during her lifetime. It is therefore likely that females of British mosquitoes normally mate only once, although little is known on this subject. In the tropical mosquito, *Aedes aegypti*, females have been shown to mate more than once. If they do not, successive egg batches are smaller and smaller (Young & Downe, 1982).

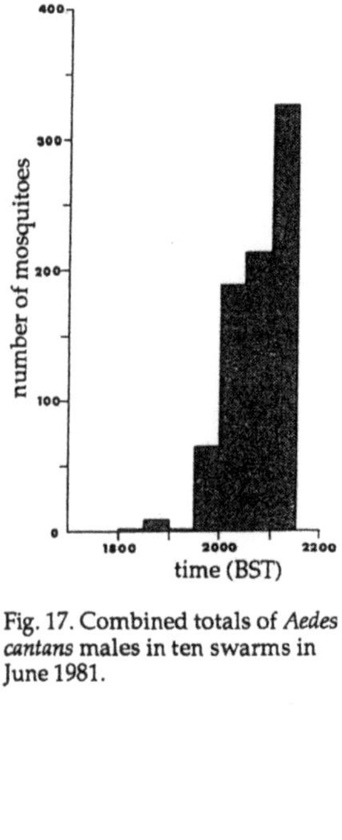

Fig. 17. Combined totals of *Aedes cantans* males in ten swarms in June 1981.

Feeding and egg development

Adults of both sexes feed on the juices of flowers and fruits. The females of most species also require a blood meal in order to obtain the nutrients necessary for the development and maturation of eggs. Such species are said to be anautogenous. A few species, namely *Culiseta subochrea* and the *molestus* form of *Culex pipiens*, can develop at least the first egg batch, and occasionally further batches, without a blood meal. These are described as autogenous. The nutrients for egg production in this case are derived

anautogenous: needing to take a blood meal before being able to lay eggs

autogenous: able to lay an egg batch without previously having fed on blood

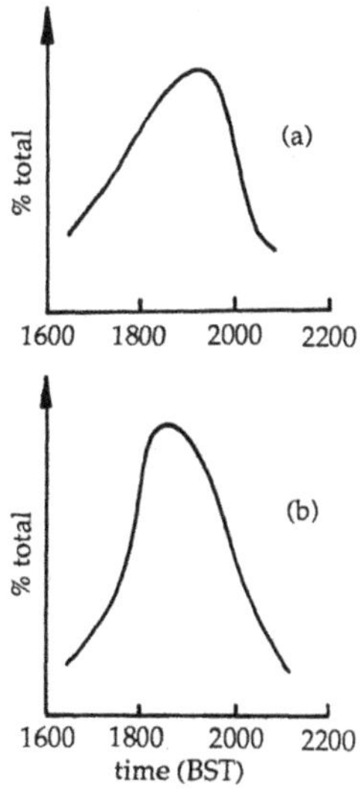

Fig. 18. Biting cycles of (a) *Aedes cantans* and (b) *Anopheles plumbeus.*

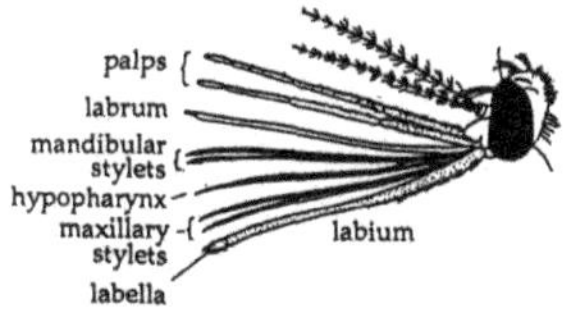

Fig. 19. Mouthparts of female anopheline, artificially displaced.

either from food reserves accumulated by the larvae or from plant juices gathered by the adult.

Mosquitoes show certain preferences for animals on which to feed (table 2). These are not necessarily dictated by their abundance. Service (1971a) showed that *Aedes caspius* and *Aedes flavescens*, from areas where rabbits, sheep, cattle and birds are common, fed only on cattle and sheep. Cattle constituted the principal hosts for *Ae. cantans, Ae. cinereus, Ae. detritus* and *Ae. punctor* collected from various sites, although a population of *Ae. cantans* from an area where cattle were absent fed on rabbits. Although many species are opportunistic and will feed whenever a host is present, they exhibit distinct peaks in activity and biting at certain times of the day. Typically they feed most commonly around dusk with variation among species (fig. 18). In *Ae. cantans* the period of maximum biting in the evening occurs 15–20 minutes after sunset (Service, 1977).

Many species feed on humans to a greater or lesser extent but in Britain none feeds on man exclusively (Service, 1968a). Thus *Ae. detritus*, which is a human-biting nuisance in many areas, also includes the blood of cattle, pigs and birds in its diet. Some human-biting species, notably *Cs. annulata* and *Cq. richiardii*, enter houses to feed on the occupants, whilst the nuisance *Aedes* species such as *Ae. cantans, Ae. punctor* and *Ae. detritus* are troublesome outdoors. *Culex pipiens*, the most widely distributed and perhaps most common species in Britain, also enters houses, often in large numbers, but does not appear to bite people.

The long, slender mouthparts of female mosquitoes are well suited for the dual functions of piercing skin and sucking blood. The conspicuous labium forms a protective sheath around the other, more delicate components (fig. 19). The skin is penetrated by the extremely fine, pointed mandibular stylets and the maxillary stylets, which are finely toothed at their tips. Another fine tube, the hypopharynx, transfers saliva from paired glands in the thorax into the skin to aid feeding. The more prominent labrum is curved inwards at its edges to form an incomplete tube. The gap in this tube is closed by the remaining

Table 2. *Host preferences of some British mosquitoes (after Service, 1971a). Figures are percentages of feeds on various hosts*

	Man	Rabbit	Horse	Cattle	Pig	Unidentified bovid	Unidentified mammal	Bird	Reptile
An. plumbeus	29.4	5.9	0	35.3	0	11.8	0	17.6	0
Ae. dorsalis	2.5	12.5	10.0	42.5	2.5	30.0	0	0	0
Cs. annulata	17.5	7.5	0	37.5	2.5	10.0	5.0	20.0	0
Cs. litorea	5.7	1.9	0	0	0	0	0	90.6	1.9
Cx. torrentium	0	8.2	0	0	0	0	0	89.8	0
Cq. richiardii	33.3	1.7	0	38.3	0	5.0	3.3	18.4	0

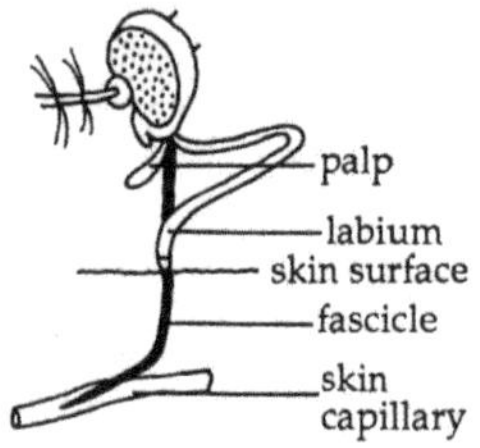

Fig. 20. Penetration of a skin blood vessel by a mosquito.

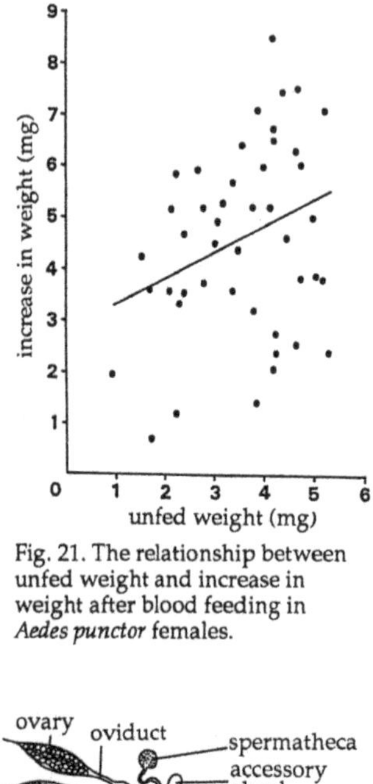

Fig. 21. The relationship between unfed weight and increase in weight after blood feeding in *Aedes punctor* females.

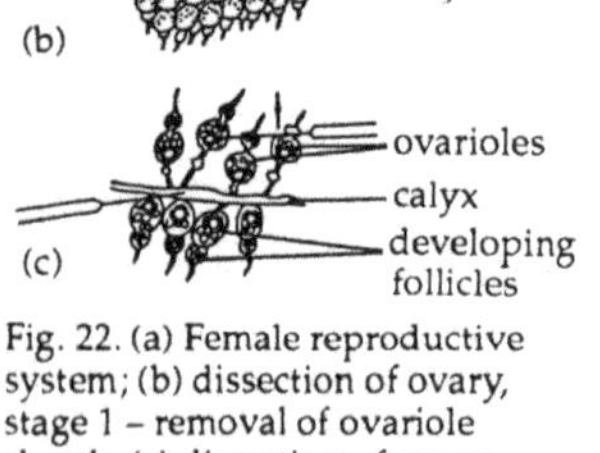

Fig. 22. (a) Female reproductive system; (b) dissection of ovary, stage 1 – removal of ovariole sheath; (c) dissection of ovary, stage 2 – isolation of ovarioles.

mouthparts to form the food canal through which blood is drawn by pumps located at the base of the food canal and in the foregut. The mouthparts of males, although basically similar in design, are not adapted for skin penetration; the mandibular and maxillary stylets are shorter.

When a female mosquito has located a suitable animal by means of odour, carbon dioxide emission, temperature and sight, it lands on the skin and probes the surface with the tip of the labrum to find a suitable area. The stylets then set to work to lacerate the skin and they enter the skin together with the labrum and hypopharynx. These mouthparts are collectively known as the fascicle. As the fascicle enters the skin the labium is bent backwards and acts as a guide (fig. 20).

By a back and forth movement the fascicle penetrates the skin deeply enough to enter or damage a capillary blood vessel so making blood available to the mosquito. Saliva flows down the hypopharynx and mixes with the blood, diluting it and making it easier to withdraw. Saliva also promotes the release of histamine from certain skin cells which causes the nearby blood capillaries to dilate. This enhances the blood flow in these vessels making available a plentiful supply of food for the female. In some, but not all, species of mosquito the saliva also contains an anticoagulant to prevent blood from clotting.

Feeding takes only a few minutes. The abdomen of the female swells and reddens as she ingests approximately her own weight of blood (Service, 1968a). Although there is no exact correlation, in general, the larger mosquitoes of a population will ingest more blood than smaller individuals (fig. 21). Similarly the number of eggs laid by an individual female is related to her size. During and after feeding many species produce a clear drop of urine, so helping to remove unwanted water from the ingested food.

In some species it has been shown that distension of the midgut (stomach) stimulates egg development. Following digestion, absorbed food products pass to the ovaries, each of which is composed of numerous subdivisions called ovarioles (fig. 22). In each reproductive cycle only one egg is matured in each ovariole. Hence the number of eggs which develop in each cycle cannot exceed the number of ovarioles.

When the eggs are fully formed the female is said to be gravid and searches for a suitable oviposition site. After egg-laying the female seeks another blood meal and the reproductive cycle of blood feeding, egg maturation and oviposition is repeated for the rest of her life.

In addition to blood, females also feed on nectar and other plant juices. If a mosquito had filled her gut with plant juices and then found a blood meal, she would then be unable to take advantage of the host. As hosts are often difficult to find this would be a clear disadvantage. In order to overcome this a mechanism exists whereby plant juices and other sugary solutions are directed into a sidebranch of the foregut called the crop, leaving the midgut empty to

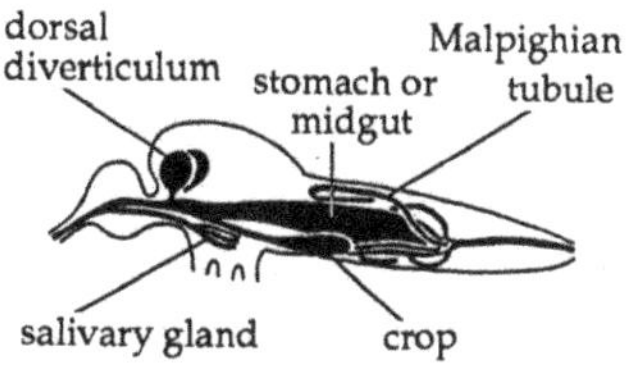

Fig. 23. Gut of a mosquito. Dissection from the side to show the major organs.

receive the next blood meal (fig. 23). At intervals small quantities of the sugary solution leave the crop to be digested and absorbed in the midgut. In some African *Culex* species, however, the sugar meal is sometimes despatched straight to the midgut and not the crop (Disney, 1970). The sources of sugar meals for different species of mosquito invite much more investigation in Britain. Little is known of the kinds of flowers visited for nectar and the importance of other sources such as honeydew and fruits.

Most species of mosquitoes in Britain have only one generation each year while some have a second and even sometimes a third. *Culex pipiens* form *molestus* shows no synchrony of development and may breed continuously throughout the year.

Overwintering

diapause: a period of suspended growth and development

British mosquitoes overwinter in various ways. Species of *Aedes* and of the subgenus *Culicella* lay eggs in the summer months in dry areas. The eggs remain in diapause until December onwards when the egg-laying sites fill with rain. The eggs then hatch and the larvae and pupae complete their development over the next few months.

Other species pass the winter as larvae. This is the case for *An. claviger*, *An. plumbeus*, *Cq. richiardii* and *Or. pulcripalpis* (Service, 1973; Fallis & Snow, 1983c). The larvae are often inactive in the colder months and develop slowly. When the water is covered with ice the larvae cannot breathe air; under such conditions they rely solely on respiration through the skin to support their limited activity and may remain buried under leaves and mud. Although larvae of *Ae. geniculatus* have been collected throughout the year they cannot survive long periods of frost. In most cases, therefore, the egg is the only overwintering stage in this as in other *Aedes* species.

Yet other species pass the winter as fertilised females. Examples are *An. atroparvus*, *An. messeae*, *Cx. pipiens*, *Cx. torrentium* and *Cs. annulata*. In the late summer and early autumn, females seek suitable places to hibernate in stables, animal shelters, lofts, attics and outbuildings. Complete hibernation occurs in the females of *An. messeae* and *Cx. pipiens* (and presumably *Cx. torrentium*, although this is not known) with individuals surviving entirely upon nutrients stored in their fat bodies. In contrast, females of *An. atroparvus* and *Cs. annulata* show incomplete hibernation, females becoming active at intervals throughout the winter to feed on the blood of humans and domestic animals (Ramsdale & Wilkes, 1985). *Cs. annulata* may also pass the winter as larvae.

With a single exception, the males of all species of British mosquitoes die as winter approaches. The exception is the *molestus* form of *Cx. pipiens* which develops in sheltered underground water, such as flooded basements of dwellings, breeding continuously and producing several generations throughout the year. In these artificial conditions *molestus* behaves unlike other temperate mosquitoes.

Larval habitats

Marshall (1938) produced an ecological classification of British mosquitoes, recognising four groups. Although some species belong in more than one category, it is of value when considering the British species. His four groups are:

1. Rural
2. Coastal
3. Arboreal
4. Domestic

The first three groups are classified according to the larval habitats used. Thus rural mosquitoes such as *Ae. cantans*, *Ae. punctor* and *Ae. rusticus* develop in ponds, pools, marshes and flooded meadows, while coastal mosquitoes such as *Ae. caspius* and *Ae. detritus* are found in brackish or saline waters. Arboreal mosquitoes such as *An. plumbeus*, *Ae. geniculatus* and *Or. pulcripalpis* develop in water-filled tree holes. Members of the fourth group are all closely associated with domestic dwellings in which they overwinter as females. Examples are *An. atroparvus*, *Cs. annulata* and *Cx. pipiens*.

Dispersal and flight range

The extent to which mosquitoes disperse from their larval habitats is extremely variable. Some woodland mosquitoes such as populations of *Ae. punctor* appear to remain in shaded woodlands in the vicinity of their temporary pools and are seldom found more than a few hundred metres from their place of emergence. At the other extreme the salt-marsh species *Ae. detritus* has been reported 6–7 km from the sea and *An. atroparvus*, another coastal species, has been found over 10 km from its larval habitats.

During the daytime adult mosquitoes rest in the shelter of vegetation, houses or animal shelters. Distinct preferences are shown; for instance *Ae. punctor* is found often among dead bracken and less frequently in grass, holly, live bracken and tree saplings. There appears to be little movement of daytime resting populations unless either the area becomes exposed to the sun or a host enters the area (Service, 1971b). During the daytime, resting mosquitoes are attracted to hosts from a distance of several metres. Little is known of night resting sites but they may well be the same as those used in the daytime.

Most mosquitoes exhibit peaks of flight activity and biting around sunset and again, to a lesser extent, at around sunrise.

3 Identification

This section includes keys to genera for all stages of British mosquitoes and to species for adults and final-instar larvae. A checklist of British species is given in table 3 (p.18). Those interested in the original descriptions of these species and the origins of the names should refer to Snow (1987a, b). Those wishing for more details and further illustrations of morphological characters should consult the keys of Cranston and others (1987) and the classic work of Marshall (1938). Since the publication of Marshall's book a number of changes in the names of our mosquitoes have been made, all of which are given in Cranston and others (1987). The important ones are that the genus *Coquillettidia* was formerly known as *Mansonia* and, before that, *Taeniorhynchus;* the genus *Culiseta* was known as *Theobaldia*; the *molestus* form of *Culex pipiens* was referred to as *Culex molestus* and *Culex territans* was called *Culex apicalis*.

Also, at the time of Marshall's publication, *Culiseta longiareolata, Culex modestus* and *Culex torrentium* had not been discovered in Britain. Although the first two of these were almost certainly wartime introductions on ships returning to our ports, *Cx. torrentium* is now known to be widely distributed and locally abundant and is replacing *Culex pipiens* in some areas of southern Britain.

Cx. modestus, Cs. longiareolata and certain other species that have not been encountered in this country for many years are included in the keys as the lack of records may well be due to oversight and not reflect their true absence.

In the future, new discoveries may well be made either of introduced species or of established species yet to be recorded. There are many species occurring in continental Europe which may become established in Britain and any specimens which cannot be identified with certainty because they do not 'key out' should be referred for further study. Such specimens, properly packaged (p. 54), may be sent to the British Mosquito Group (address on p. 63).

Use of keys

For eggs, use Key I; for pupae, use Key III. For larvae, use Key II as a quick guide to genus and then work through Key V to identify the species. For adults, use Key IV as a quick guide to genus and then work through Key VI to identify the species.

Table 3. *A list of mosquito species recorded in Britain*

Subfamily Anophelinae

Anopheles

Anopheles (Anopheles) algeriensis Theobald, 1903
Anopheles (Anopheles) atroparvus van Thiel, 1927
Anopheles (Anopheles) claviger (Meigen, 1804)
Anopheles (Anopheles) messeae Falleroni, 1926
Anopheles (Anopheles) plumbeus Stephens, 1828

Subfamily Culicinae

Aedes

Aedes (Aedes) cinereus (Meigen, 1818)
Aedes (Aedimorphus) vexans (Meigen, 1830)
Aedes (Finlaya) geniculatus (Olivier, 1791)
Aedes (Ochlerotatus) annulipes (Meigen, 1830)
Aedes (Ochlerotatus) cantans (Meigen, 1830)
Aedes (Ochlerotatus) caspius (Pallas, 1771)
Aedes (Ochlerotatus) communis (De Geer, 1776)
Aedes (Ochlerotatus) detritus (Haliday, 1833)
Aedes (Ochlerotatus) dorsalis (Meigen, 1830)
Aedes (Ochlerotatus) flavescens (Müller, 1764)
Aedes (Ochlerotatus) leucomelas (Meigen, 1804)
Aedes (Ochlerotatus) punctor (Kirby, 1837)
Aedes (Ochlerotatus) rusticus (Rossi, 1790)
Aedes (Ochlerotatus) sticticus (Meigen, 1838)

Coquillettidia

Coquillettidia (Coquillettidia) richiardii (Ficalbi, 1889)

Culex

Culex (Barraudius) modestus Ficalbi, 1889
Culex (Culex) pipiens Linnaeus, 1758
(inc. form *molestus* Forskål, 1775)
Culex (Culex) torrentium Martini, 1925
Culex (Neoculex) territans Walker, 1856

Culiseta

Culiseta (Allotheobaldia) longiareolata (Macquart, 1838)
Culiseta (Culicella) fumipennis (Stephens, 1825)
Culiseta (Culicella) litorea (Shute, 1928)
Culiseta (Culicella) morsitans (Theobald, 1901)
Culiseta (Culiseta) alaskaensis (Ludlow, 1906)
Culiseta (Culiseta) annulata (Schrank, 1776)
Culiseta (Culiseta) subochrea (Edwards, 1921)

Orthopodomyia

Orthopodomyia (Orthopodomyia) pulcripalpis (Rondani, 1872)

Key I. Eggs: key to genera

I.1

I.2

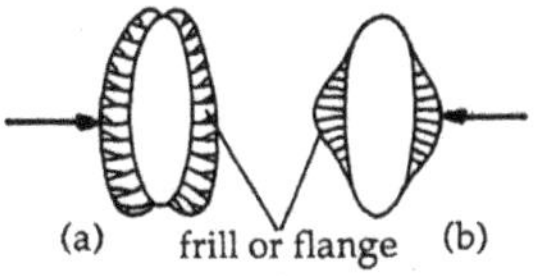

I.3

I.4

I.5

I.6

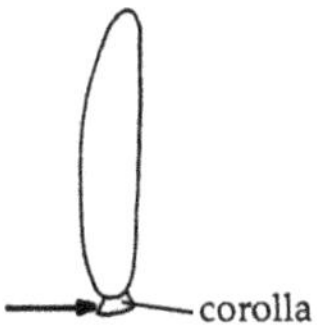

I.7

1 Eggs laid singly or in small groups (I.1), on water or in dry places subject to flooding 2

– Eggs laid in rafts on water (I.2) 5

2 With frill or flange running part or all the way around the egg (I.3) 3

– No such frill or flange (I.4) 4

3 Flange with branched ribs; without floats at side; egg round in cross section (I.3a) *Orthopodomyia*

– Flange with unbranched ribs forming floats at the sides; egg not round in cross-section (I.3b) *Anopheles*

4 Shape as in I.5 or broader (one side usually flattened) *Aedes*

– Shape as in I.6 or narrower (one end rounded and one end pointed) *Culiseta*, subgenus *Culicella*

5 With disc-like corolla at lower end (I.7) *Culex*

– Corolla absent 6

6 Outline with numerous small projections seen at x100 magnification (I.8) *Coquillettidia*

– Outline almost completely smooth at x100 magnification (I.9) *Culiseta*, subgenus *Culiseta*

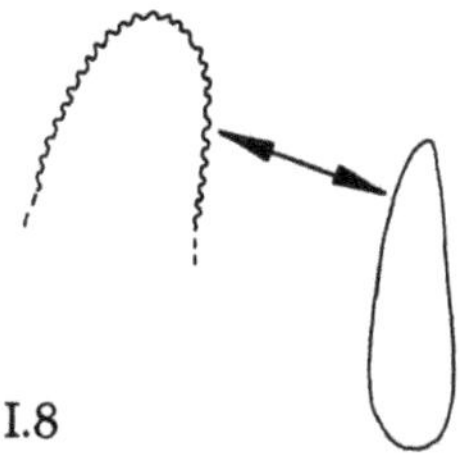

I.8

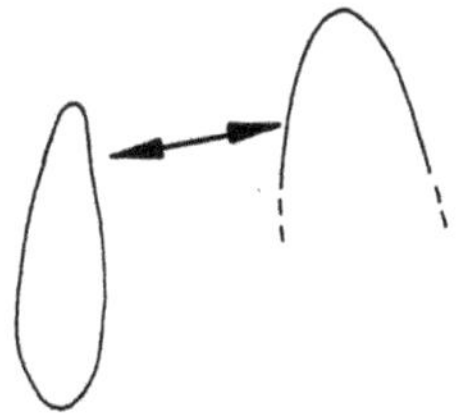

I.9

Key II. Larvae: key to genera

Introductory note

Larval features are illustrated and named in fig. 24 (p. 20). The eighth abdominal segment bears a *respiratory siphon* (fig. 4, p. 4) in all genera apart from *Anopheles* (fig. 3, p. 4). Except in *Orthopodomyia pulcripalpis* and *Coquillettidia richiardii*, the siphon is equipped with two rows of spines; each row is called a *pecten* (fig. 6, p. 5). In *Anopheles* there is no siphon and the spiracles open on the surface of segment 8 within a *spiracular plate*. The pectens of *Anopheles* lie each side of the spiracular plate and each takes the form of a row of spines arising from one side of a triangular plate (fig. 4, p. 4).

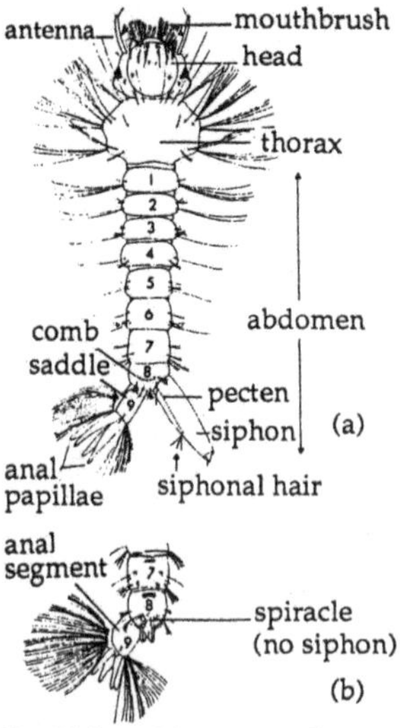

Fig. 24. Larval features. (a) Larva of any genus except *Anopheles*; (b) *Anopheles* larva.

1 Respiratory siphon present on abdominal segment 8 (II.1) — 2

– Respiratory siphon absent (II.2) — *Anopheles*

2 Siphon modified for piercing aquatic plants (II.3) *Coquillettidia* (*Cq. richiardii* is the only British species)

– Siphon not so modified (II.4) — 3

3 Siphon without pecten spines (II.5); antennae smooth, without minute spines (II.6); abdominal segments 7 and 8 with dorsal plates; body pink in life — *Orthopodomyia* (*Or. pulcripalpis* is the only British species)

– Siphon with two basal rows of pecten spines (II.7); antennae usually with minute spines (II.8); abdominal segments 7 and 8 without dorsal plates; body not pink — 4

4 Siphon with at least three pairs of hairs (II.9) (each hair usually has more than one branch) — *Culex*

– Siphon with only one pair of hairs (II.10) — 5

5 Siphon with paired hairs inserted near the base in the second to fourth instars (II.11) and inserted near the middle in the first instar; antennal hair of first instar unbranched (II.12) — *Culiseta*

Note. First-instar larvae are easily recognised by the possession of an egg breaker on the top of the head (fig. 3, p. 4)

– Siphon with paired hairs inserted near the middle in all instars (II.13); antennal hair of first instar branched, and often three-branched (II.14) — *Aedes*

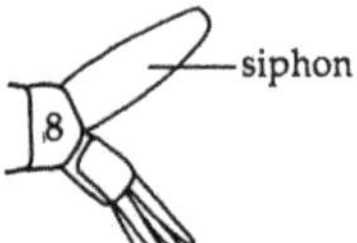

II.1

II.2

II.3

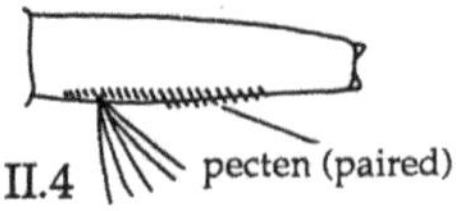

II.4

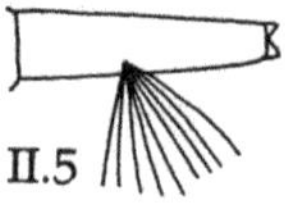

II.5

II.6

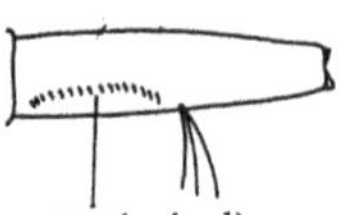

II.7

II.8

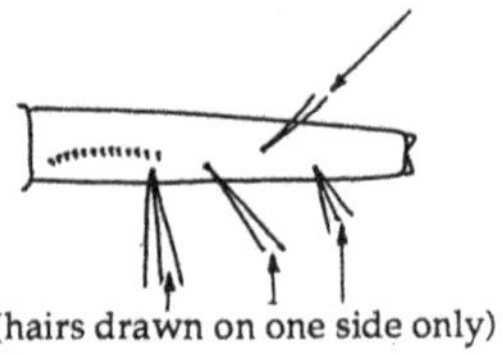

II.9

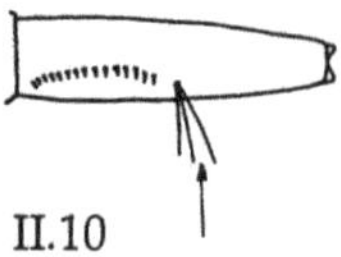

II.10

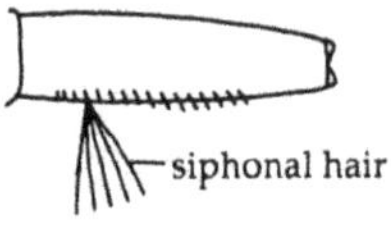

II.11

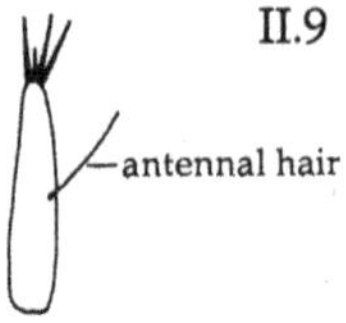

II.12

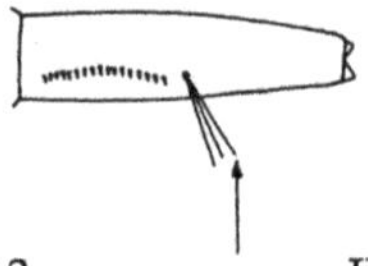

II.13

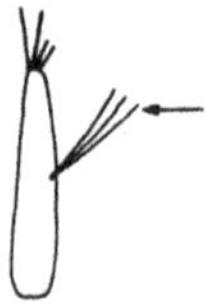

II.14

Key III. Pupae: key to genera

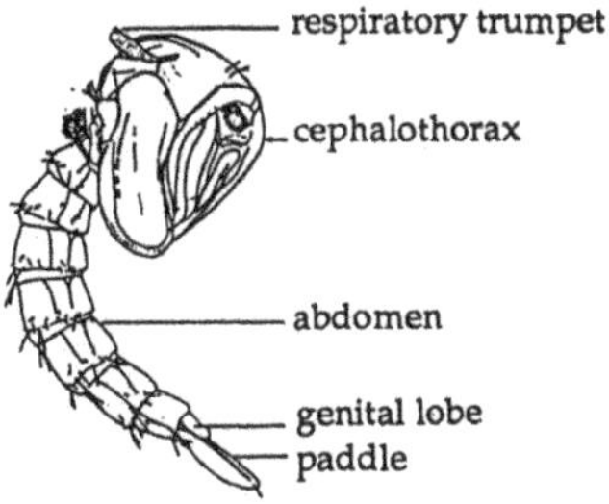

Fig. 25. Pupal features.

Introductory note

Pupal features are illustrated and named in fig. 25. The bulbous cephalothorax (fused head and thorax) bears a pair of *respiratory trumpets* which in *Coquillettidia richiardii* are modified to pierce aquatic plants. In all the British species they are tubular with an opening which is split in *Anopheles* but unsplit in the remaining genera (fig. 8, p. 5). The abdomen is composed of nine segments, each with characteristic hairs, especially hair 9 on most segments (fig. 9, p. 5). From the last segment arises a pair of paddles equipped with one or two pairs of hairs. If only one pair of hairs is present these are termed primary paddle hairs (1–P) while if a second pair is present they are called accessory paddle hairs (2–P).

III.1

III.2

1 Respiratory trumpets modified for piercing aquatic plant tissue (III.1); paddles narrow and with a notch at the tip (III.2) — *Coquillettidia*

– Respiratory trumpets not modified (III.3); paddles broader and without a notch (III.4) — 2

III.3

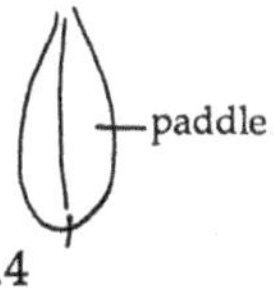

III.4

2 Paddles with accessory hair (2–P) (III.5) — 3

– Paddles lacking accessory hair — 4

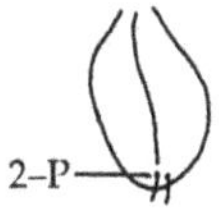

III.5

3 Accessory paddle hair (2–P) on lower (ventral) surface and in front of primary paddle hair (1–P) (III.6); hair 9 on segments 3–7 a spine (III.7) — *Anopheles*

– Accessory paddle hair (2–P) on upper (dorsal) surface and at side of primary paddle hair (1–P) (III.8); hair 9 on segments 3–7 not a spine (III.9) — *Culex*

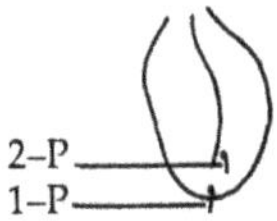

III.6

4 Paddle shape as in III.10; hair 9 on segment 8 about half length of paddle and strongly branched; dorsal abdominal hairs short — *Orthopodomyia*

– Paddle shape otherwise; hair 9 on segment 8 much less than half paddle length (III.11); dorsal abdominal hairs well developed — *Aedes* and *Culiseta*

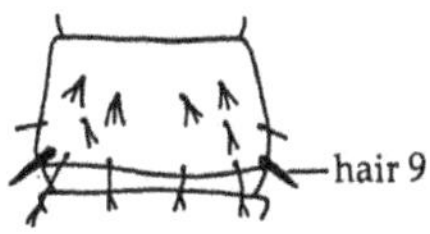

III.7

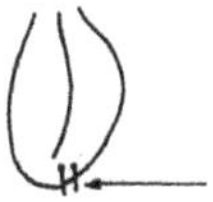

III.8

III.9

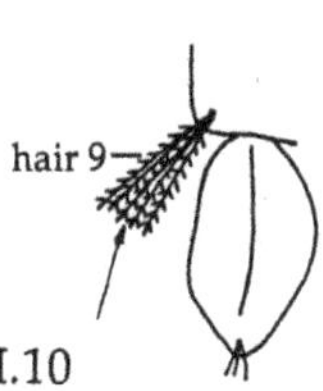

III.10

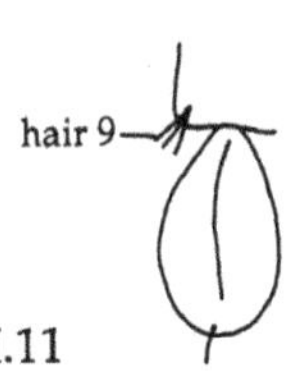

III.11

Key IV. Adults: key to genera

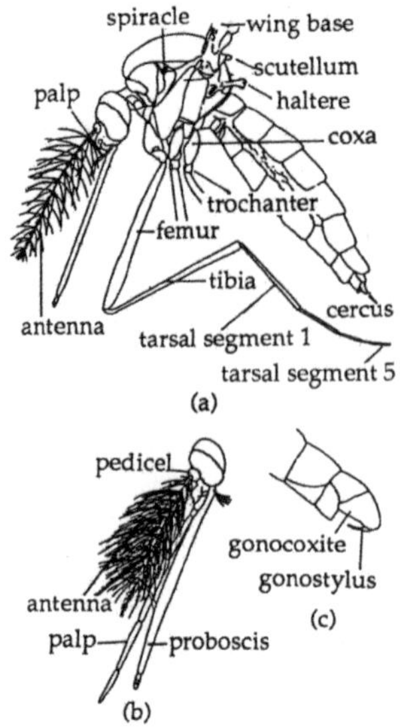

Fig. 26. Adult features. (a) Female; (b) head of male; (c) rear of male.

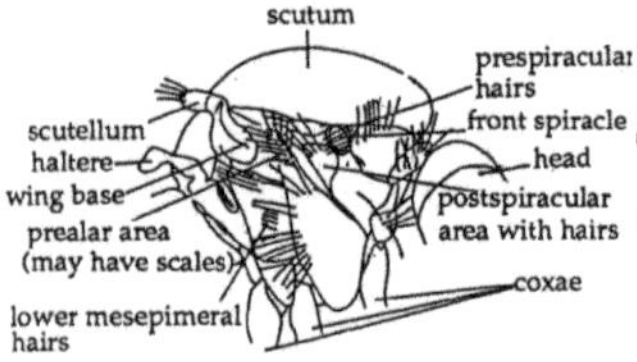

Fig. 27. Thorax of mosquito from the side.

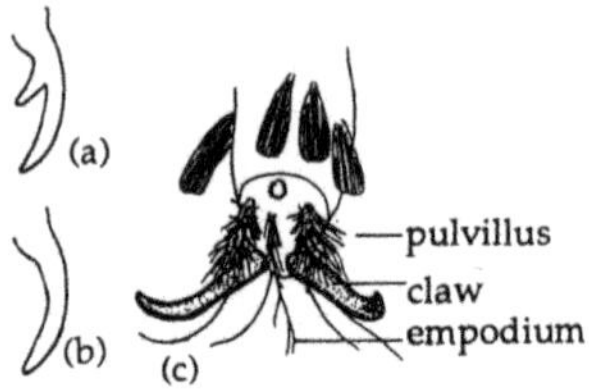

Fig. 28. Foot claw (a) with inner tooth and (b) without inner tooth. (c) Foot of *Culex pipiens* showing claws, empodium and pulvilli.

Introductory note

Adult features are illustrated and named in fig. 26. Before attempting to identify an adult mosquito it is necessary to determine the sex. Male mosquitoes are distinguished from females by several characters, the most obvious of which is that in males the antennae have whorls of long hairs (*plumose antennae*) whereas in females the hairs on the antennae are short (*pilose antennae*) (fig. 13, p. 7).

In the following key, characters refer to both males and females unless otherwise stated.

The mouthparts comprise a long *proboscis*, on either side of which are the *maxillary palps*, often referred to simply as *palps*.

The top of the thorax is roofed by a large plate called the *scutum* behind which is a small *scutellum*. On the side of the thorax are two pairs of *spiracles*, front and rear, and several groups of scales and hairs (fig. 27).

The three pairs of legs are long and each is composed of a *coxa* (plural *coxae*), *trochanter* (plural *trochanters*), *femur* (plural *femora*), *tibia* (plural *tibiae*) and *tarsus* (plural *tarsi*) of five segments (fig. 26). The feet end in claws which are toothed in some species (fig. 28). A spine (*empodium*) and, in *Culex*, a pair of pad-like *pulvilli* (singular *pulvillus*) also arise from the end of each foot.

In females the abdomen ends in a pair of *cerci* (singular *cercus*).

1 Palps of female no longer than half proboscis length (IV.1); abdomen covered with scales; scutellum three-lobed behind (IV.2) — 2

– Palps of female about as long as proboscis (IV.3); abdomen with hairs but very few scales; scutellum rounded behind (IV.4) — *Anopheles*

2 Fore- and mid-leg claws of female with inner tooth (IV.5); female abdomen evenly tapered to tip; female cerci long and easily seen from above (IV.6); postspiracular hairs present (IV.7) — *Aedes*

– Fore- and mid-leg claws of female lacking inner tooth (IV.8); female abdomen parallel-sided, rounded at tip; female cerci short, hardly visible from above (IV.9); postspiracular hairs absent — 3

3 Tarsi without rings of pale scales but with pulvilli (IV.10); in male, palps upturned (IV.11) — *Culex*

– Tarsi with rings of pale scales; feet without pulvilli; in male, palps not upturned — 4

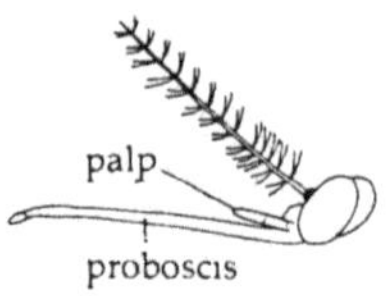

IV.1

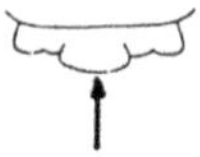

IV.2

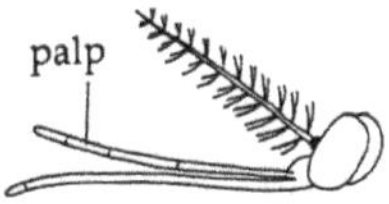

IV.3

IV.4

4 Tarsi mainly dark with rings of pale scales on segment 1 of mid legs and all segments of hind legs (IV.12); top of thorax black with white stripes (IV.13); female palps about half proboscis length (IV.14) *Orthopodomyia*

– Rings of pale scales on tarsal segments of all legs; top of thorax coloured otherwise; female palps no more than one-third proboscis length (IV.15) 5

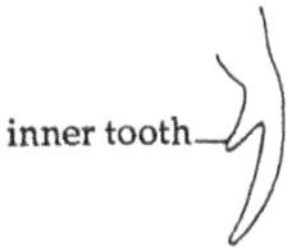

IV.5

5 Prespiracular hairs present (IV.16); wing vein scales narrow (IV.17) *Culiseta*

– Prespiracular hairs absent; wing vein scales broad (IV.18) *Coquillettidia*

IV.6

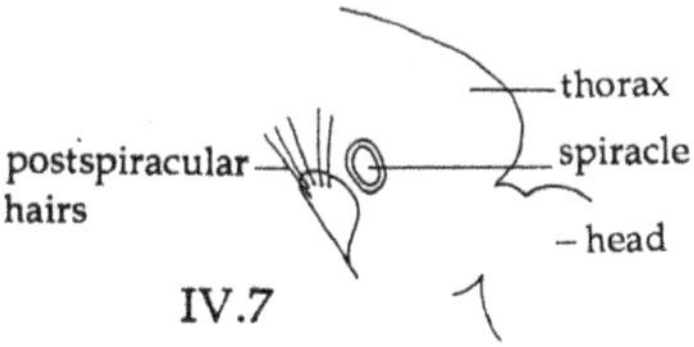

IV.7

IV.8

IV.9

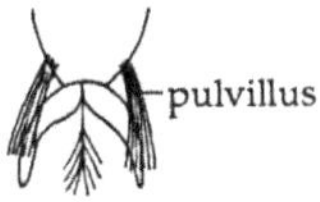

IV.10

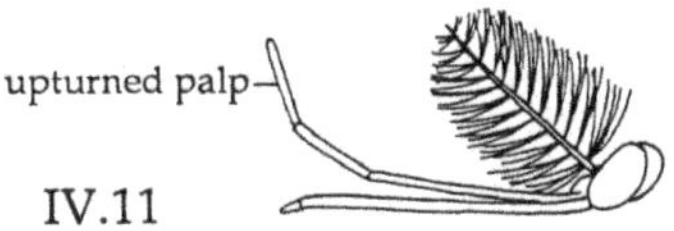

IV.11

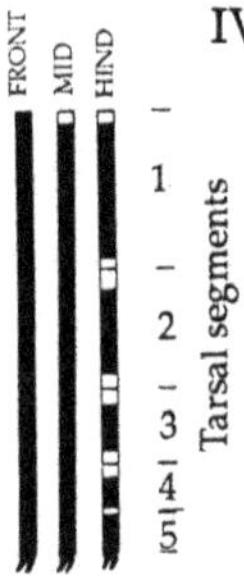

IV.12

IV.13

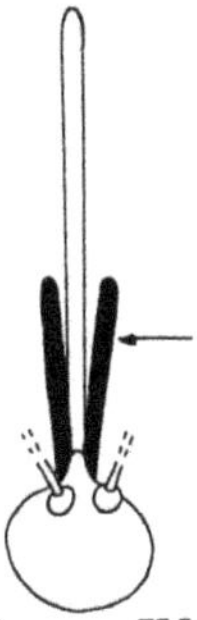

IV.14

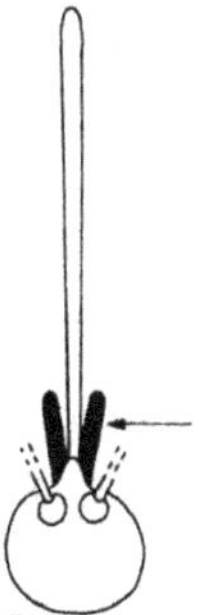

IV.15

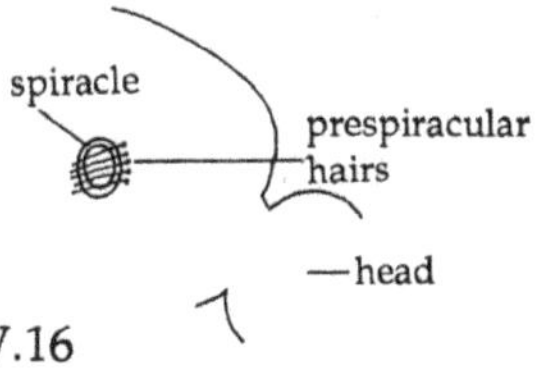

IV.16

IV.17

IV.18

Key V. Fourth-instar larvae: key to species

Introductory note

Larval features are illustrated and named in fig. 1 (p. vi) and fig. 24 (p. 20). The larval *mouthbrushes* (fig. 16, p. 11) are complex and difficult to examine but the design of the inner hairs of the mouthbrushes separates some species and points to their feeding habits. Thus larvae such as *Aedes cinereus* with smooth hairs are filter feeders while those such as *Aedes punctor* with hairs which have saw-like edges also browse.

The eighth abdominal segment bears a *respiratory siphon* (fig. 4, p. 4) in all genera apart from *Anopheles* (fig. 3, p. 4). Except in *Orthopodomyia pulcripalpis* and *Coquillettidia richiardii*, the siphon is equipped with two rows of spines; each row is called a *pecten* (fig. 6, p. 5). In *Anopheles* there is no siphon and the spiracles open on the surface of segment 8 within a *spiracular plate*. The pectens of *Anopheles* lie each side of the spiracular plate and each takes the form of a row of spines arising from one side of a triangular plate (fig. 4, p. 4). The siphon bears 1–5 pairs of *siphonal hairs* which are often branched.

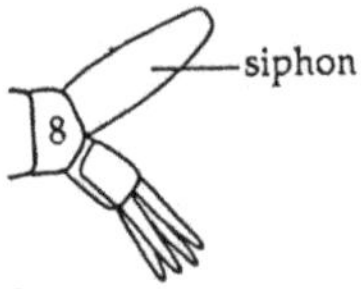

V.1

V.2

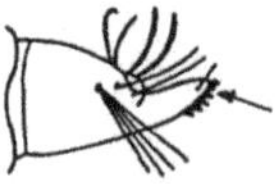

V.3

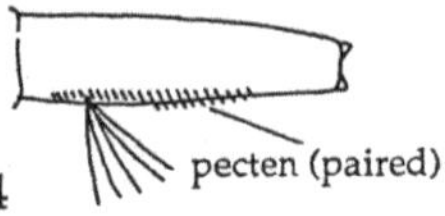

V.4

It is often necessary to consider the *siphonal index* which is calculated by dividing the length of the siphon by its width (fig. 4, p. 4).

On each side of the eighth abdominal segment of culicine larvae and first instar *Anopheles* larvae there is a group of *comb scales* forming the paired *combs*. These scales are not present in second- to fourth-instar *Anopheles* larvae.

The last abdominal segment, often called the anal segment, ends in four *anal papillae*. It also bears the prominent *caudal brush* and *ventral brush*. Most or all of the hairs of the ventral brush arise from a barred area and these are termed *cratal hairs*. Those hairs originating outside the barred area are called *precratal hairs* (fig. 4, p. 4). The anal segment is either partly or fully encased by a plate called the *saddle* from which arises a pair of *saddle hairs* (1–X).

Fourth-instar larvae are distinguished from earlier instars by their size and the complexity of body hairs, pectens and combs. This requires both experience and judgement. If in doubt, it is advisable to keep larvae until they pupate and examine the cast larval skins.

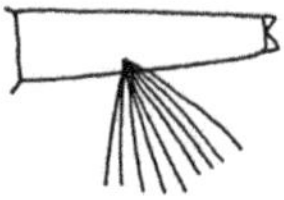

V.5

1 Respiratory siphon present on abdominal segment 8 (V.1) subfamily Culicinae 2

– Respiratory siphon absent (V.2) subfamily Anophelinae 24

V.6

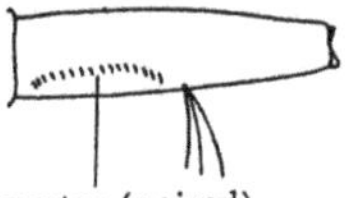

V.7

2 Siphon modified for piercing aquatic plants (V.3) *Coquillettidia richiardii*

– Siphon not so modified (V.4) 3

3 Siphon without pecten spines (V.5); antenna smooth, without minute spines (V.6); abdominal segments 7 and 8 with dorsal plates; body colour pink in life *Orthopodomyia pulcripalpis*

– Siphon with two basal rows of pecten spines (V.7); antenna usually with minute spines (V.8); abdominal segments 7 and 8 without dorsal plates; body not pink 4

V.8

4 Siphon with at least three pairs of hairs (V.9) genus *Culex* 5

– Siphon with only one pair of hairs (V.10) 8

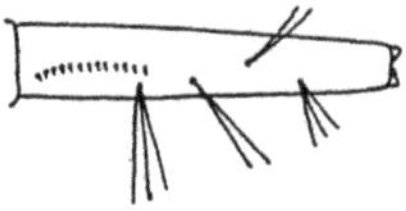

(hairs drawn on one side only)

V.9

5 Siphon long and slender; Siphonal Index (siphon length divided by width) 6 or more (V.11); cuticle of thorax and abdomen covered with minute spines *Culex territans*

– Siphonal Index less than 6 (V.12); body cuticle without minute spines 6

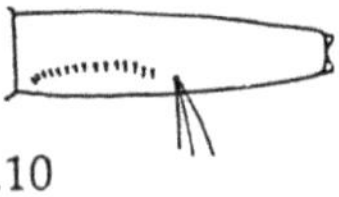

V.10

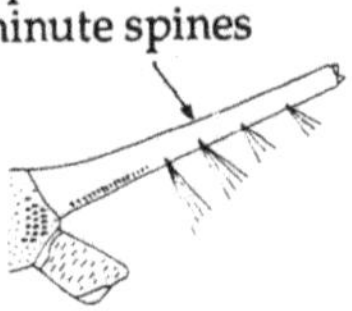

V.11

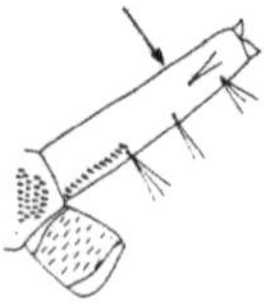

V.12

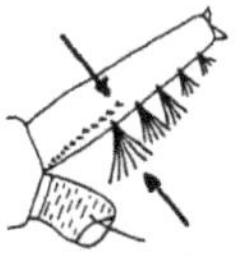

V.13
(hairs drawn on one side only)

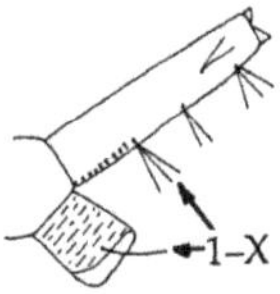

V.14

6 Siphon with three to five pairs of hairs, some of which arise before the end of the row of pecten spines (V.13); the most basal hair with at least six branches *Culex modestus*

– Siphon with four or five pairs of hairs, all arising beyond the row of pecten spines (V.14); the most basal hair with up to four branches 7

7 Length of thoracic hair 1–T normally greater than one-half the length of thoracic hair 2–T; hair 1–X (V.14) and hair 3–I usually double with 3–I sometimes triple (V.15) *Culex torrentium*

– Length of thoracic hair 1–T normally less than one-half the length of 2–T; hair 1–X and 3–I usually, but not always, single; 3–I never triple (V.15) *Culex pipiens* (including form *molestus*)

Note. Unless 3–I is triple it is impossible to separate these two species with certainty as larvae.

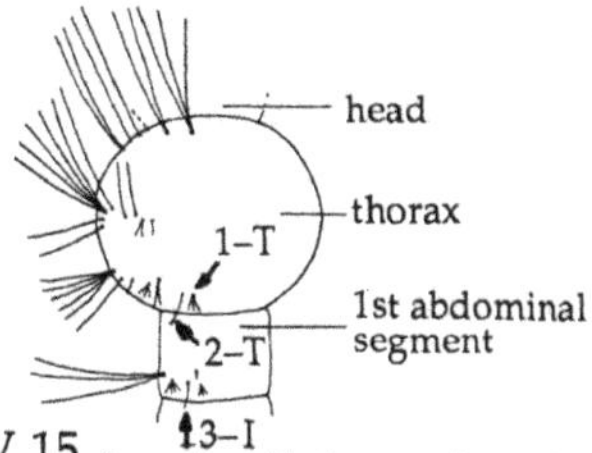

V.15 (lengths of hairs not shown)

8 Siphon with paired hairs arising basally (V.16) genus *Culiseta* 9

– Siphon with paired hairs arising approximately midway between base and tip (V.17) genus *Aedes* 13

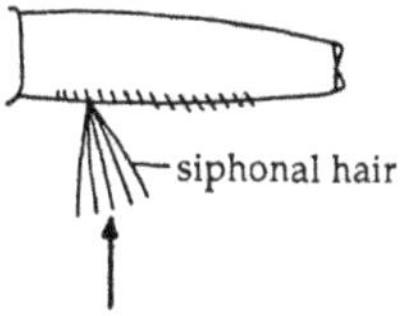

V.16

9 Pecten with some spines hair-like (V.18) *Culiseta alaskaensis* and *Culiseta annulata* and *Culiseta subochrea*

Note. These three species are indistinguishable as larvae.

– Pecten composed entirely of strong spines; no hair-like spines present (V.19) 10

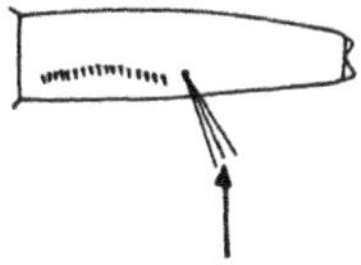

V.17

V.18

V.19

10 Antenna smooth, appreciably shorter than head (V.20); Siphonal Index (siphon length divided by width) less than 3 (V.21) *Culiseta longiareolata*

– Antenna with small spines and longer than head (V.22); Siphonal Index more than 4 (V.23) subgenus *Culicella* 11

11 Tip of siphon with a pair of large feathery hairs; pecten extending beyond middle of siphon with the last few spines widely spaced (V.24) *Culiseta fumipennis*

– Tip of siphon with a pair of minute unbranched hairs; pecten confined to basal one-third of siphon (V.25) 12

V.20

V.21

V.22

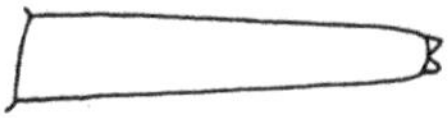

V.23

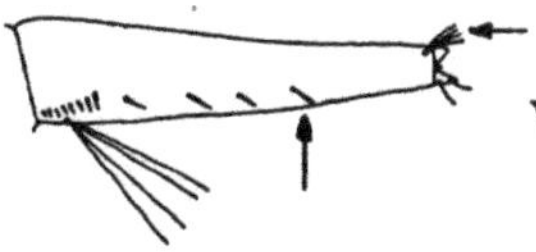

V.24

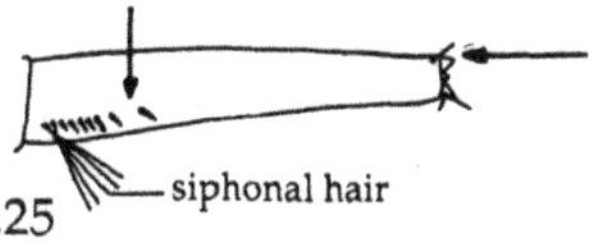

V.25

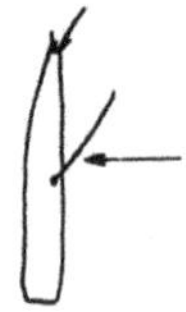
V.26

12 Siphonal hair usually less than 0.4 times length of siphon *Culiseta morsitans*

– Siphonal hair usually more than 0.4 times length of siphon *Culiseta litorea*

Note. It may not be possible to distinguish these two species as larvae.

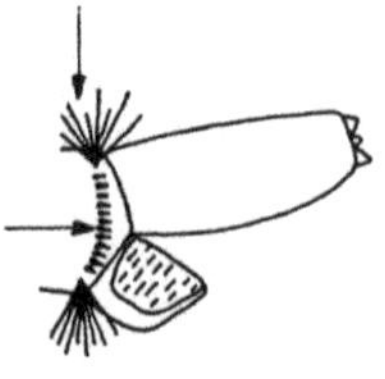
V.27

13 Antenna smooth; antennal hair unbranched (V.26); abdominal segments 1–8 with star-shaped hairs (V.27); comb scales in an even row (V.27) *Aedes geniculatus*

– Antenna with minute spines; antennal hair branched (V.28); hairs on abdominal segments 1–8 not star-shaped; comb scales arranged in a patch or uneven row (V.29) 14

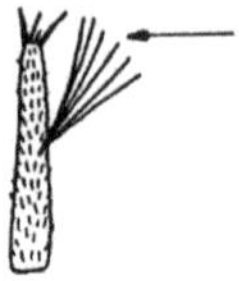
V.28

14 Siphon with three or four pairs of dorsal hairs; pecten usually with 22–28 spines, the end one or two more widely spaced and extending beyond the siphonal hair (V.30) *Aedes rusticus*

– Siphon without dorsal hairs; pecten usually with 12–30 spines, none beyond the siphonal hair (V.31) 15

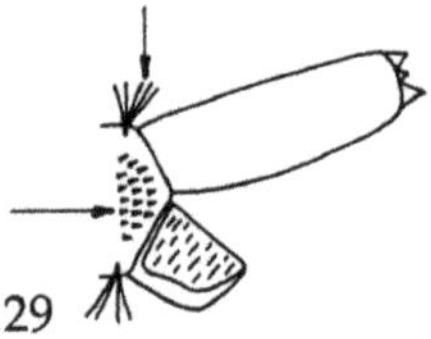
V.29

15 Last one or two pecten spines widely spaced (V.32) 16

– Last spines of pecten not widely spaced (V.33) 17

16 Mouthbrushes composed exclusively of smooth hairs (V.34); anal papillae nearly as long as siphon (V.35) *Aedes cinereus*

– Tips of inner hairs of mouthbrushes toothed like a saw (V.36); anal papillae usually much shorter than siphon (V.37) *Aedes vexans*

V.30

17 Anal papillae about three times the length of the saddle *Aedes sticticus*

– Anal papillae less than twice the length of the saddle 18

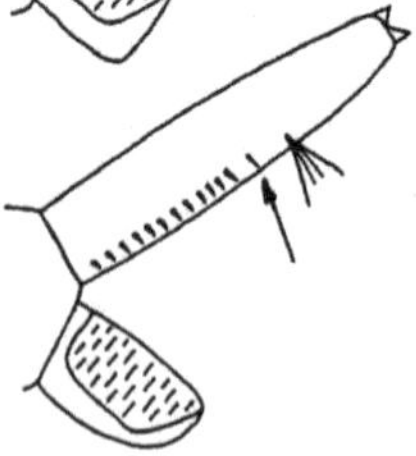
V.31

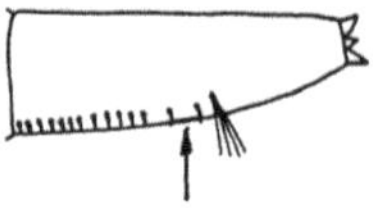
V.32

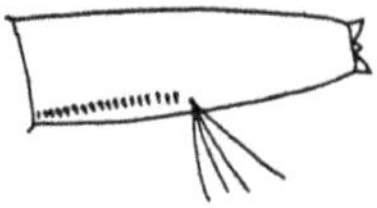
V.33

V.34

V.36

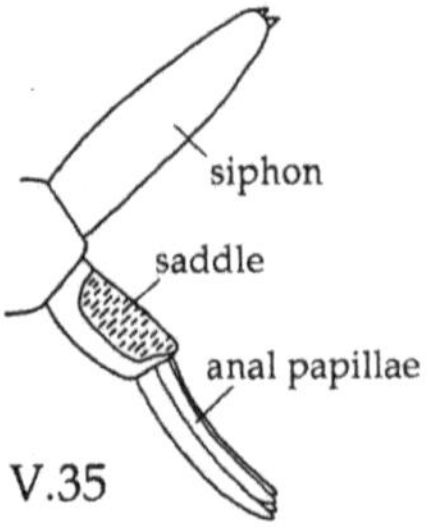

V.35

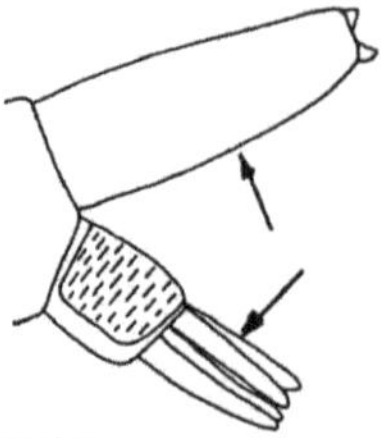
V.37

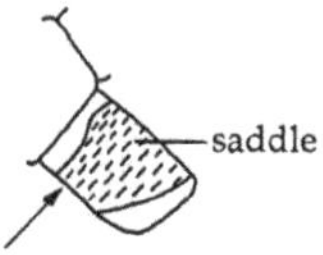

V.38

18 Saddle completely encircles the anal segment (V.38) *Aedes punctor*

– Saddle does not completely encircle the anal segment (V.39) 19

19 At least 45 comb scales, usually more 20

– Fewer than 45 comb scales, usually fewer than 35 21

V.39

20 Anal papillae no more than half saddle length (V.40) *Aedes detritus*

– Anal papillae at least as long as saddle, often longer (V.41) *Aedes communis*

21 One or two precratal hairs (V.42) *Aedes leucomelas* and *Aedes caspius* and *Aedes dorsalis*

Note. These three species are indistinguishable as larvae.

– At least three precratal hairs 22

V.40

22 At least six precratal hairs; Siphonal Index (siphon length divided by width) greater than 3; saddle hair (1–X) at least 1.25 times as long as saddle (V.42) *Aedes flavescens*

– No more than five precratal hairs; Siphonal Index less than 3; saddle hair no longer than saddle 23

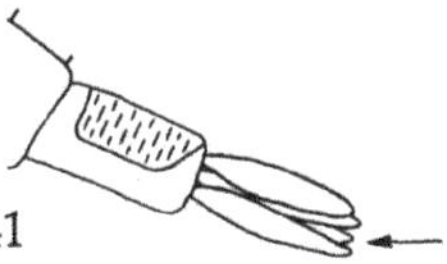

V.41

23 Normally 28–38 comb scales (in shaded woodland pools) *Aedes cantans*

– Normally 31–44 comb scales (in open, sunlit pools in heathland or meadows) *Aedes annulipes*

Note. Comb-scale numbers overlap and larvae may not be distinguishable morphologically.

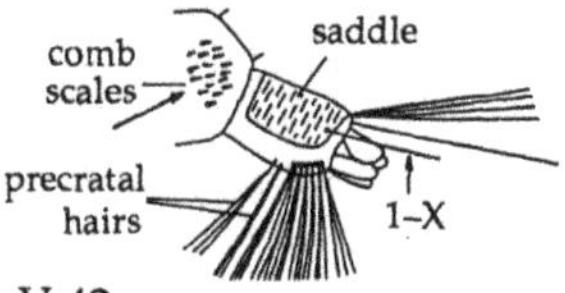

V.42

24 Frontal hairs (5–, 6– and 7–C) of head small and unbranched (V.43a) *Anopheles plumbeus*

– Frontal hairs large and feathery (V.43b) 25

25 Outer clypeal hairs (3–C) (V.43a) with multiple branches (V.44) *Anopheles atroparvus* and *Anopheles messeae*

– Outer clypeal hairs unbranched or with few branches (V.45) 26

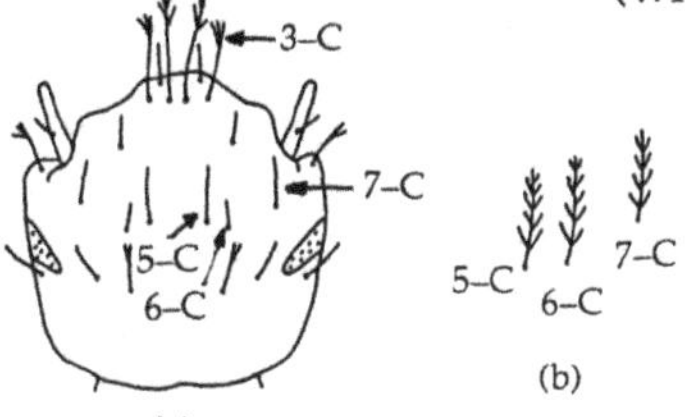

V.43

V.44

V.45

26 Head with extensive pattern of dark marks (V.46); antenna with numerous minute spines (V.46); origin of saddle hair well within margin of saddle (V.47); more than one short spine in each gap between the longest spines of the pecten (V.48) *Anopheles algeriensis*

– Head with less extensive markings (V.49); antenna with few minute spines (V.49); origin of saddle hair just outside edge of saddle (V.50); most of the gaps between the long pecten spines contain only one short spine (V.51) *Anopheles claviger*

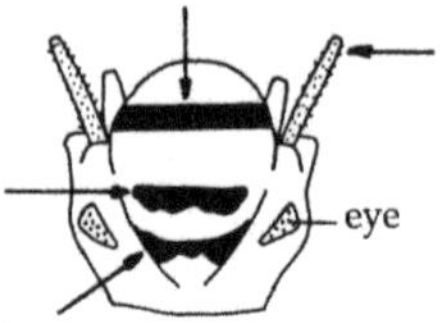

V.46

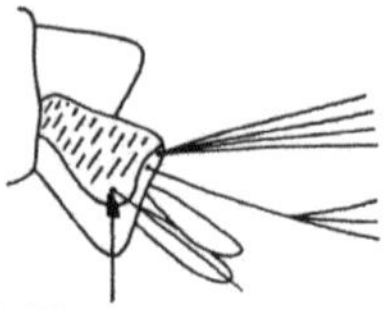
V.47

V.48

V.49

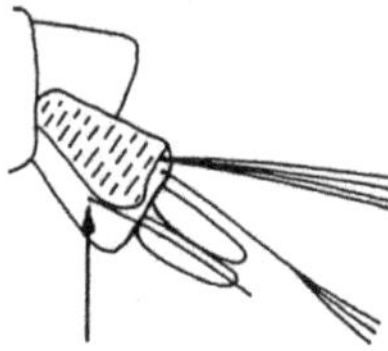
V.50

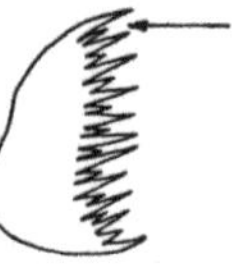
V.51

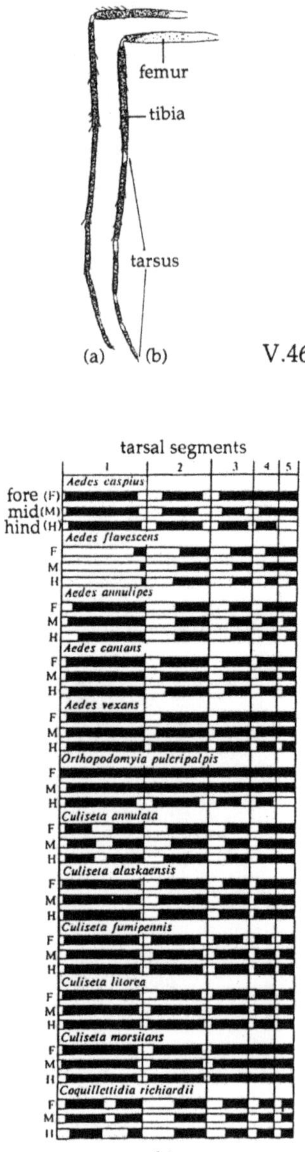

Fig. 29. Tarsal patterns (a) without rings (*Aedes geniculatus*) and (b) with rings below joints. (c) Table of tarsal band patterns of selected species.

Key VI. Adults: key to species

Introductory note

Adult features are illustrated and named in figs. 1, 26 and 27 (p. 22). Before attempting to identify an adult mosquito it is necessary to determine the sex. Male mosquitoes are distinguished from females by several characters, the most obvious of which is that in males the antennae have whorls of long hairs (*plumose antennae*) whereas in females the hairs on the antennae are short (*pilose antennae*; fig. 13, p. 7).

In the following key, characters refer to both males and females unless otherwise stated.

The mouthparts comprise a long *proboscis*, on either side of which are the *maxillary palps*, often referred to simply as *palps*. The top of the thorax is roofed by a large plate called the *scutum* behind which is a small *scutellum*. On the side of the thorax are two pairs of *spiracles*, front and rear, and several groups of scales and hairs (fig. 27, p. 22).

The three pairs of legs are long and each is composed of a *coxa* (plural *coxae*), *trochanter* (plural *trochanters*), *femur* (plural *femora*), *tibia* (plural *tibiae*) and *tarsus* (plural *tarsi*) of five segments (fig. 29). The feet end in *claws* which are toothed in some species (fig. 28, p. 22). A spine (*empodium*) and, in *Culex*, a pair of pad-like *pulvilli* (singular *pulvillus*) also arise from the end of each foot.

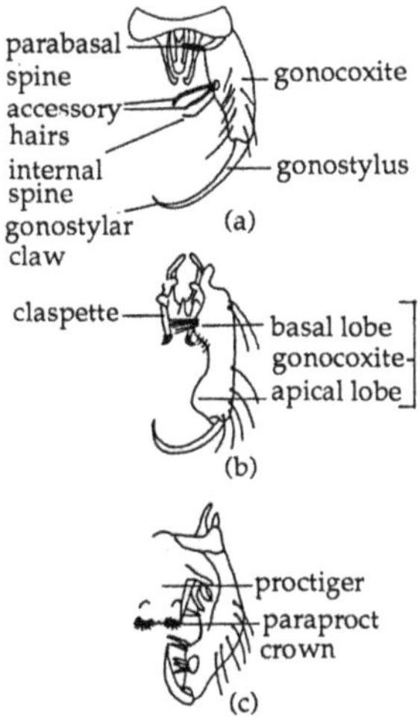

Fig. 30. Hypopygium (male external genitalia) of (a) *Anopheles*, (b) *Aedes* and (c) *Culex* (only one side of each is drawn).

In females the abdomen ends in a pair of *cerci* (singular *cercus)* but in males the abdomen terminates in the *hypopygium* (plural *hypopygia)*, or external genitalia (fig. 30). It is important to realise that the hypopygium rotates through 180° during the 12–24 hours following emergence and this may lead to confusion regarding which surface is above and which below. Before examining newly emerged male specimens it is advisable to leave these to mature for at least 24 hours. When hypopygia are illustrated in the key they are shown in mature males viewed from below. Techniques for preparing male genitalia for examination are described in chapter 5 (p. 54).

Certain features of the male hypopygium are used in the keys and need further explanation. The hypopygium consists of a central *aedeagus* (penis) surrounded by a *phallosome* and the whole flanked by a pair of *gonocoxites*. Each gonocoxite bears a *gonostylus* which in turn bears a *gonostylar claw* (fig. 30). The gonocoxite may have projections, the main one being the *basal lobe* which is best developed in *Aedes*. In *Anopheles* there is instead a *parabasal lobe* bearing stout *parabasal spines* (fig. 30). Also of note in *Anopheles* is a group of spines, about mid-way along the inner face of the gonocoxite, which includes the *internal spine* (fig. 30) referred to in the key.

Between the bases of the gonocoxites in *Aedes* and *Anopheles* are a pair of elongated *claspettes*, each comprising a *stem* bearing a *filament*.

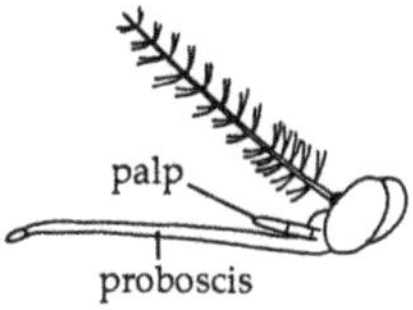

VI.1

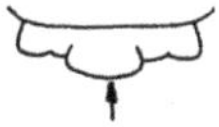

VI.2

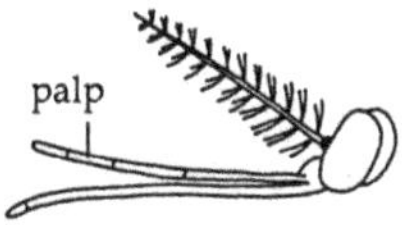

VI.3

VI.4

1 Palps of female not more than half proboscis length (VI.1); abdomen covered with scales; scutellum three-lobed behind (VI.2) subfamily Culicinae 2

– Palps of female about as long as proboscis (VI.3); abdomen with hairs but very few scales; scutellum rounded behind (VI.4) subfamily Anophelinae 28

2 Fore- and mid-leg claws of female with inner tooth (VI.5); female abdomen evenly tapered to tip; cerci long and visible from above (VI.6); postspiracular hairs present in both sexes (VI.7) genus *Aedes* 3

– Fore- and mid-leg claws of female lacking inner tooth (VI.8); female abdomen parallel-sided, rounded at tip; cerci short, hardly visible from above (VI.9); postspiracular hairs absent in both sexes 16

3 Tarsi without rings of pale scales 4

– Tarsi with rings of pale scales (VI.10) 11

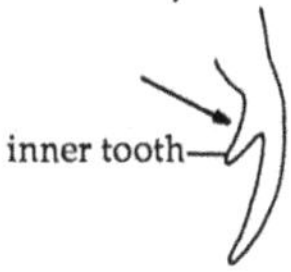

VI.5

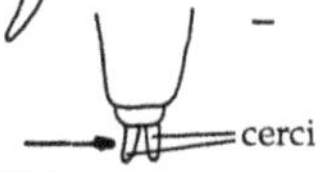

VI.6

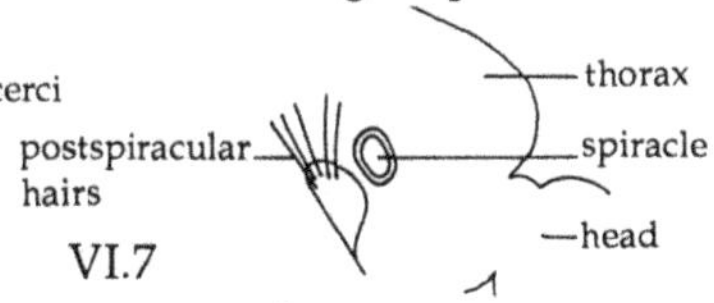

VI.7

VI.8

VI.9

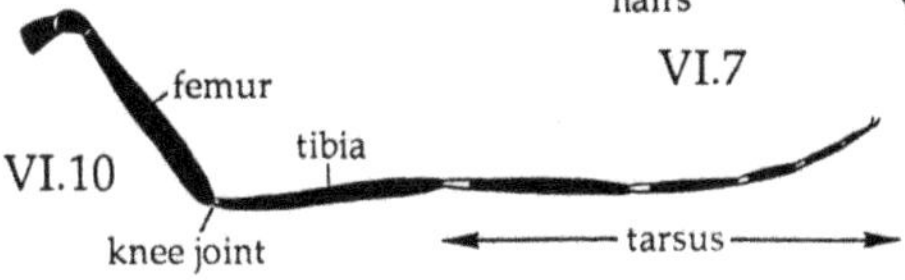

VI.10

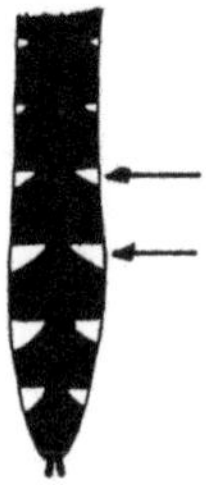
VI.11

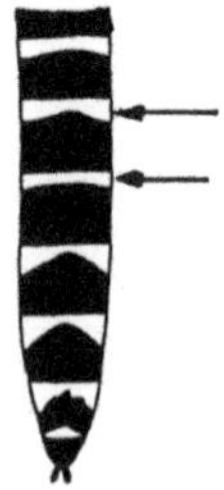
VI.12

4 Patches of pale scales at sides of abdomen do not join over the top of the abdomen but are separated by an area of dark scales in the midline (VI.11) 5

– Patches of pale scales at sides join, resulting in complete pale bands across the abdomen (VI.12) 6

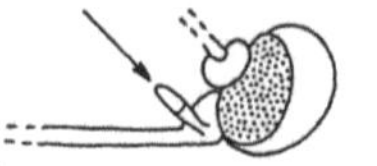
VI.13

5 Scutum reddish-brown with scattered pale scales; pale scales at sides of abdomen form two longitudinal stripes which are not easily visible from above; pale scales on 'knee joints' (tips of femora) inconspicuous; male palps short, same length as in female (VI.13) *Aedes cinereus* (p. 38)

– Scutum black with pale scales forming a distinctive pattern; pale scales at sides of abdomen arranged as conspicuous triangular patches, readily seen from above (VI.11); pale scales on 'knee joints' conspicuous (VI.14); male palps only slightly shorter than the proboscis *Aedes geniculatus* (p.38, pl.3)

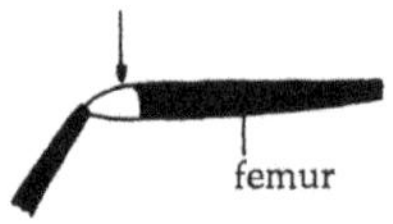

VI.14

6 Abdominal pale bands of female widened centrally to form a broken stripe along the middle of the upper side of the abdomen, especially visible towards the rear (VI.15); stripe not evident in male; female palp at least a quarter proboscis length; last segment of male palp distinctly swollen (VI.16a); male hypopygium with basal lobe of gonocoxite bearing a group of flattened sword-like hairs and with a wavy gonostylar claw (VI.17a) *Aedes rusticus* (p. 42)

– Abdominal pale bands of female not broadened centrally and not forming a stripe; female palp less than one-quarter proboscis length; last segment of male palp not distinctly swollen (except in *Ae. punctor* to an extent) (VI.16b); male hypopygium with basal lobe of gonocoxite with spines but not flattened, sword-like hairs; gonostylar claw straight (VI.17b) 7

VI.15

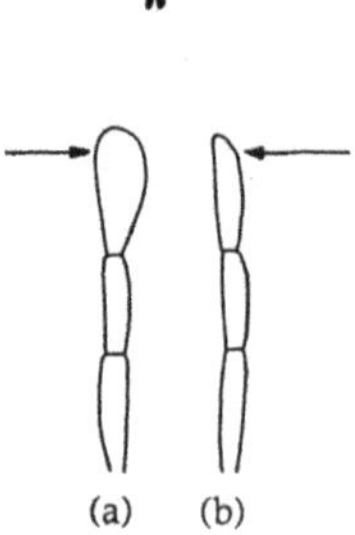

VI.16

7 Lower mesepimeral hairs absent; hind tibiae with whitish stripe on outer side (not always present in male); scutum dark brown in the centre and mostly white-scaled at sides *Aedes sticticus* (p. 42)

– Lower mesepimeral hairs present (VI.18); hind tibiae without whitish stripe but with scattered pale scales; scutum either uniformly brown or a mixture of shades of brown 8

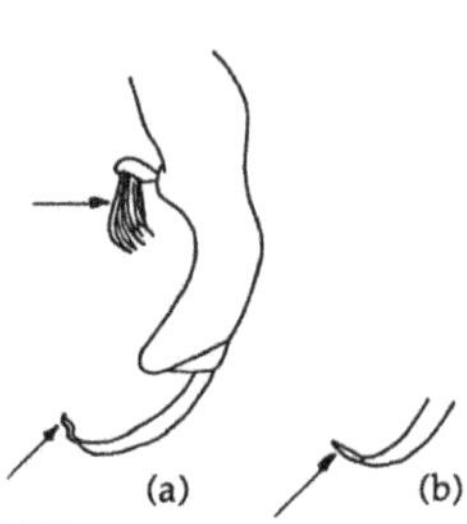

VI.17

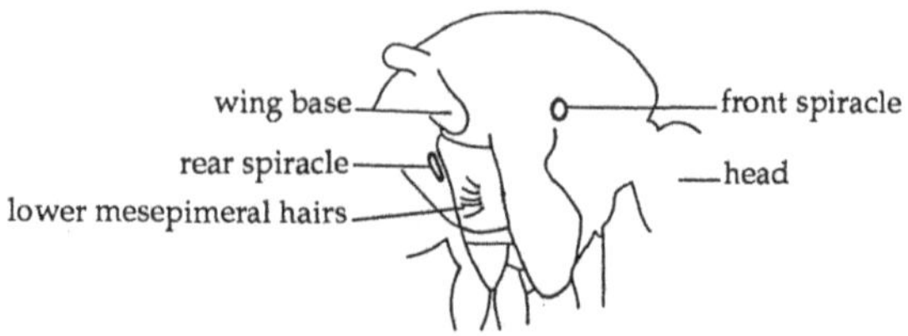

VI.18

PLATE 1

1. *Anopheles plumbeus*
2. *Anopheles maculipennis* (*atroparvus / messeae*)

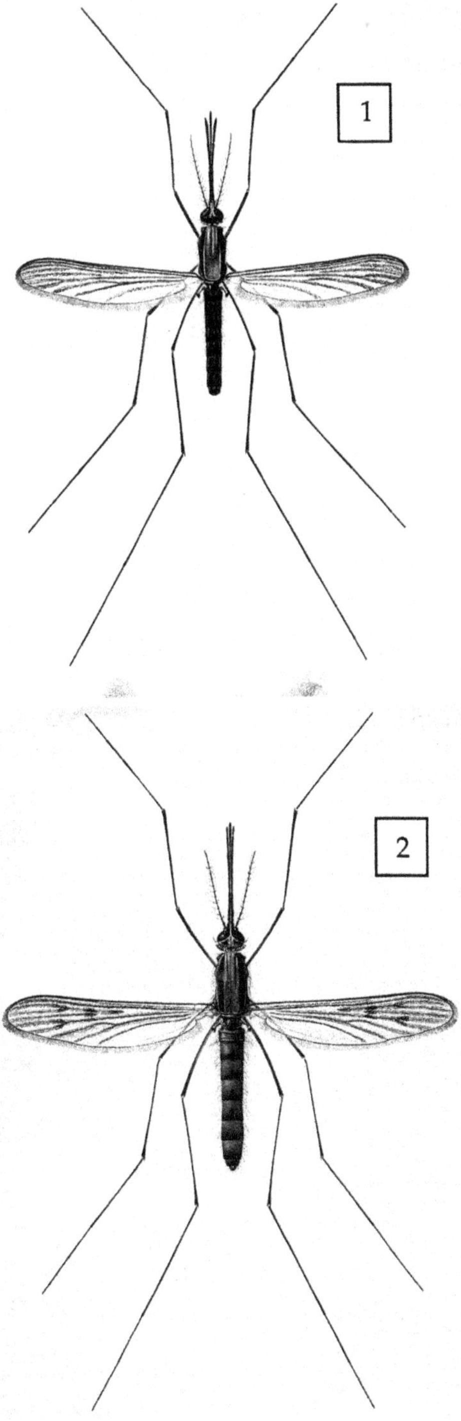

The mosquitoes in plates 1 to 4 are all female

PLATE 2

1. *Culex pipiens*
2. *Culiseta annulata*

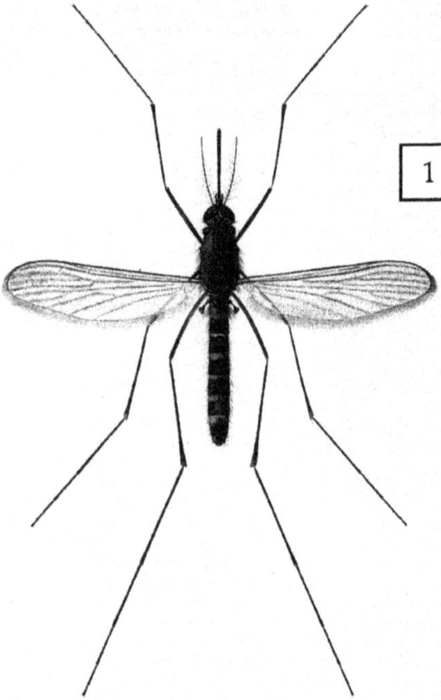

1

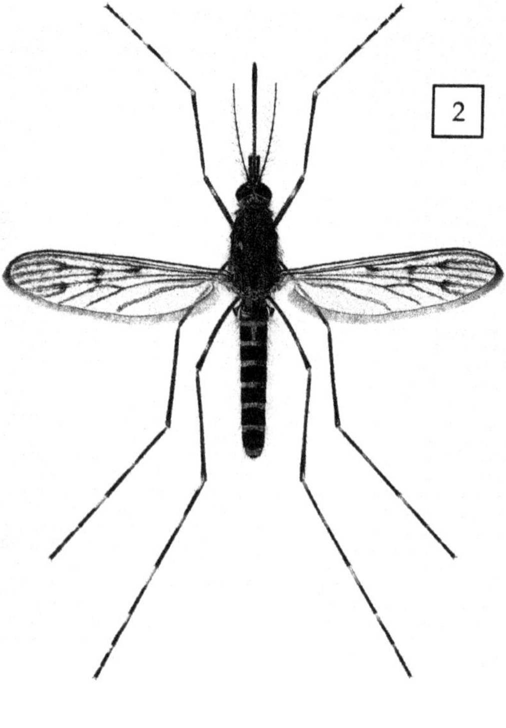

2

PLATE 3

1. *Aedes geniculatus*
2. *Aedes detritus*

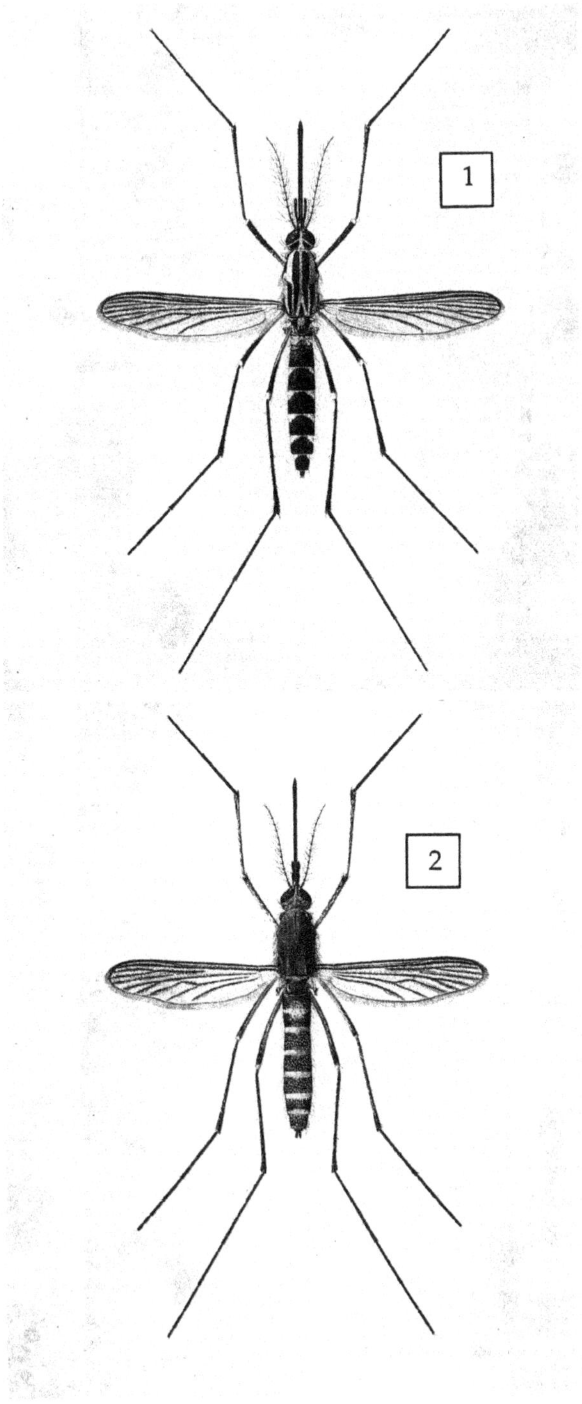

PLATE 4

1. *Coquillettidia richiardii*

2. *Orthopodomyia pulcripalpis*

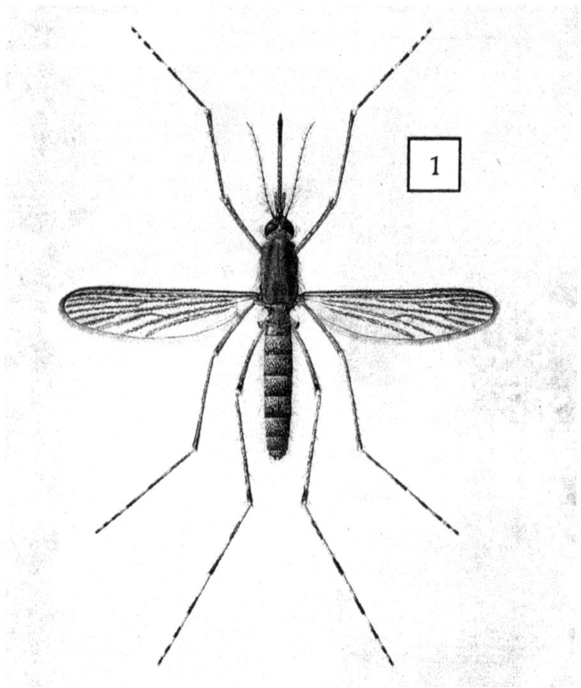

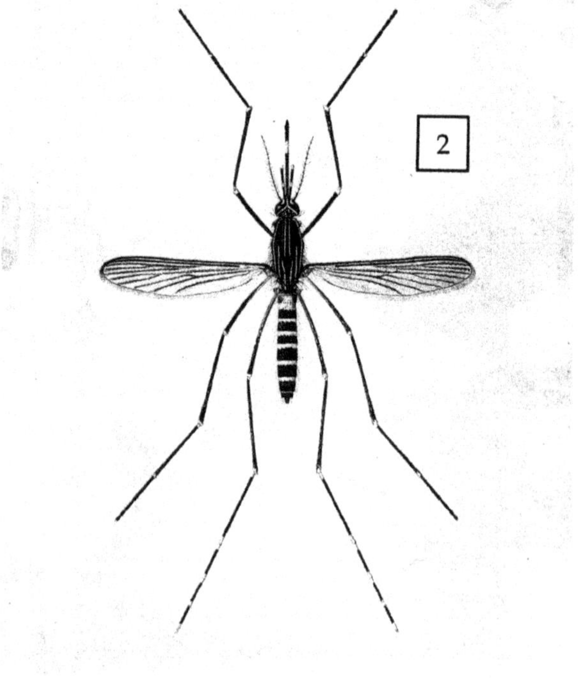

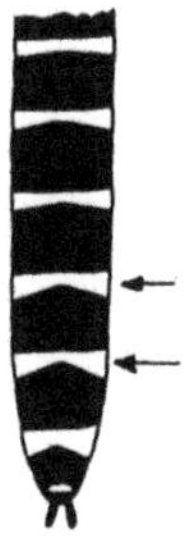

VI.19

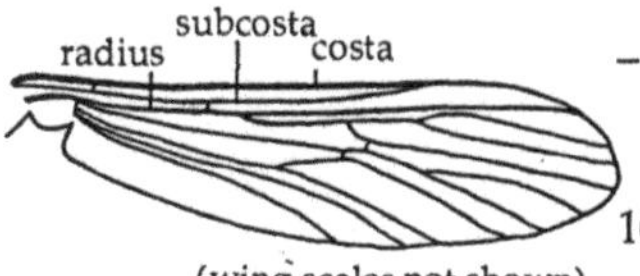

VI.20 (wing scales not shown)

8 Abdominal pale bands narrowed centrally, especially on rear segments (VI.19); wings entirely dark-scaled *Aedes punctor* (p. 41)

– Abdominal pale bands of more or less uniform width; wings with pale scales present, at least on bases of costa, subcosta and radius (VI.20) and perhaps extending into outer half of wing 9

9 Pale scales of wing veins restricted to bases of costa, subcosta and radius; front of fore and mid tibiae with few scattered pale scales *Aedes communis* (p. 40)

– Pale scales of wing veins scattered on most veins; front of fore and mid tibiae with numerous pale scales 10

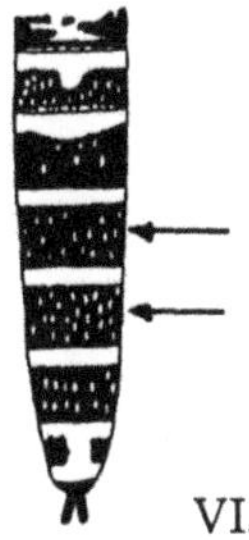

VI.21

VI.22

10 First segment of antenna with white scales below; a small patch of pale scales present immediately below the front spiracle (VI.18); dark areas of abdomen with few or no scattered pale scales; male palps with numerous pale scales on next-to-last segment *Aedes leucomelas* (p. 41)

– First segment of antenna lacking pale scales below; no patch of pale scales present immediately below the front spiracle; dark areas of abdomen with numerous scattered pale scales (VI.21); male palps usually completely dark-scaled *Aedes detritus* (p. 40, pl. 3)

11 Tarsi with pale rings both above and below the joints, surrounding joints (VI.22); pale bands on abdomen extended centrally to produce a more-or-less continuous longitudinal stripe (VI.23) 12

– Tarsi with pale rings below the joints only (VI.24); when present, pale bands on abdomen not broadened centrally (VI.25) 13

VI.23

VI.24

12 Scutum with middle third reddish-brown with two narrow white stripes behind, the two outer thirds mainly off-white but with two dark longitudinal stripes behind (VI.26); wing veins R_S and M (VI.27) mainly pale-scaled; abdomen black with white markings *Aedes dorsalis* (p. 41)

– Scutum reddish-fawn with two white stripes running from front border to hind margin and two further pairs of pale stripes within rear area (VI.28); wing veins R_S and M with an even mixture of light and dark scales; abdomen brown with yellowish markings *Aedes caspius* (p. 39)

VI.25

VI.26

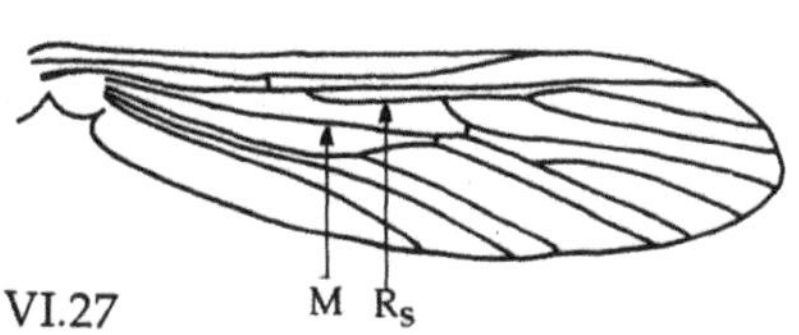

VI.27

VI.28

VI.29

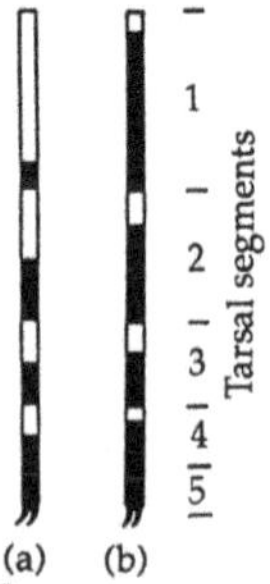

VI.30

13 Abdomen mainly yellowish with a varying number of black scales which may form dark stripes in the middle and on each side (VI.29) and which are more conspicuous in males than females; first tarsal segments pale-scaled except for a narrow dark ring at the tip (VI.30a) *Aedes flavescens* (p. 41)

– Abdomen with pale bands above contrasting with dark areas (VI.31); first tarsal segments with pale scales forming basal band (VI.30b) 14

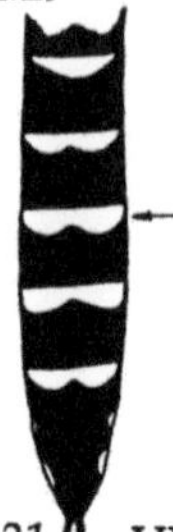

VI.31

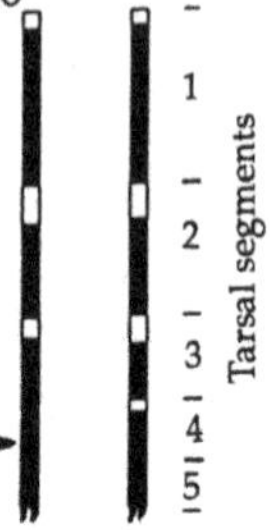

VI.32

14 In female, pale abdominal bands distinctly two-lobed (VI.31) (more or less uniform width in male); scales of wing veins and proboscis predominantly dark; rings of pale scales on first to third segments of fore tarsus but absent on fourth and fifth segments (VI.32a); male hypopygium with enlarged gonostylus and with gonostylar claw originating just before the tip (VI.33a) *Aedes vexans* (p. 38)

– In male and female, pale abdominal bands of more or less uniform width; scales of wing veins and proboscis both dark and pale, mixed; rings of pale scales present on first to fourth segments of fore tarsus (VI.32b); male hypopygium with narrow gonostylus and gonostylar claw originating at the tip (VI.33b) 15

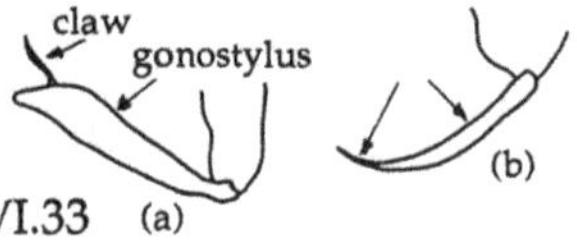

VI.33

15 All pale scales yellowish, dark scales dark brown; pale rings on second to fifth segments of hind tarsi each occupying about half of the segment (VI.34a); male hypopygium with basal lobe of gonocoxite small, two-lobed and without a bristle; stem of claspette as wide as filament, filament rectangular (VI.35) *Aedes annulipes* (p. 39)

– All pale scales white, dark scales blackish-brown; pale rings on second to fifth segments of hind tarsi each occupying about one third of the segment (VI.34b); male hypopygium with basal lobe of gonocoxite long and prominent and bearing a bristle; stem of claspette much narrower than filament, filament triangular (VI.36) *Aedes cantans* (p. 39)

Note. Separation of females of these two species can be extremely difficult, especially with worn, older specimens.

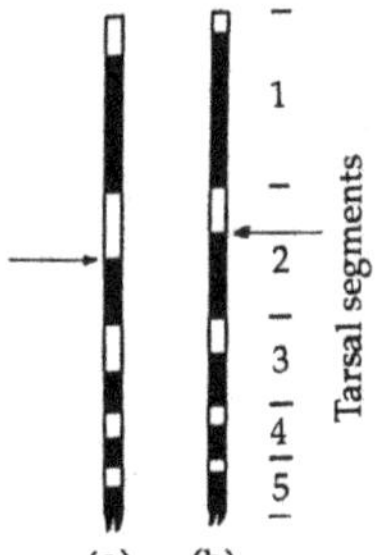

VI.34

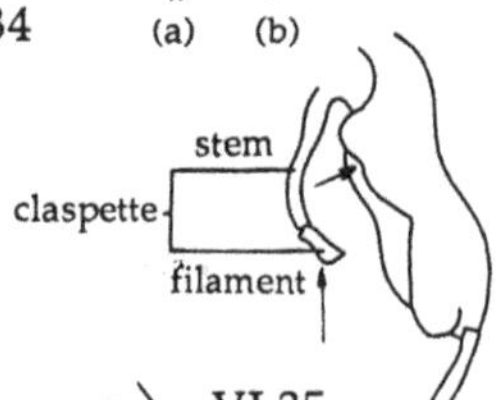

VI.35

VI.36

16 Tarsi without rings of pale scales; legs with pulvilli (VI.37); male palps upturned (VI.38) genus *Culex* 17

– Tarsi with rings of pale scales; legs without pulvilli; male palps not upturned 20

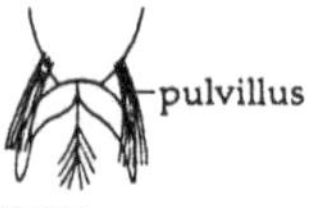

VI.37

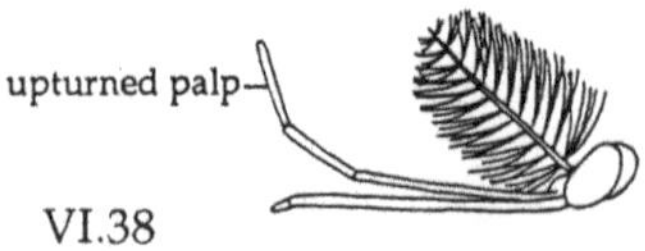

VI.38

17 Abdominal segments without pale bands above (VI.39); patches of pale scales at sides sometimes extending into stripes; first segment of hind tarsus distinctly shorter than hind tibia (VI.40a); male palps with only a few short spines (VI.41a) *Culex modestus* (p. 43)

– Abdominal segments with bands of pale scales above (VI.42) extending to form large patches at sides; first segment of hind tarsus about equal to, or longer than, hind tibia (VI.40b); male palps with tufts of long hairs (VI.41b) 18

18 Transverse bands of pale scales on hind margins of abdominal segments (viewed from above) (VI.43) *Culex territans* (p. 44)

– Transverse bands of pale scales on front margins of abdominal segments (VI.44) 19

19 Prealar scales normally present (VI.45), often four or more; fore tibia with inconspicuous pale spot above; male hypopygium as in VI.46 *Culex torrentium* (p. 44)

– Prealar scales usually absent; if present never more than two; fore tibia without pale spot; male hypopygium as in VI.47

Culex pipiens (including form *molestus*) (p. 43, pl. 2)

Note. It is not possible to distinguish the *molestus* form from the typical form of *pipiens* morphologically.

20 Tarsi mainly dark with rings of pale scales on first tarsal segment of middle leg and all hind tarsal segments (VI.48); scutum black with white stripes (VI.49); female palps about half proboscis length (VI.50) *Orthopodomyia pulcripalpis* (p. 47, pl. 4)

– Rings of pale scales on tarsal segments of all legs; scutum ornamented otherwise; female palps no more than one third proboscis length (VI.51) 21

21 Prespiracular hairs absent; wing vein scales broad (VI.52a) *Coquillettidia richiardii* (p. 42, pl. 4)

– Prespiracular hairs present (VI.53); wing vein scales narrow (VI.52b) genus *Culiseta* 22

VI.39

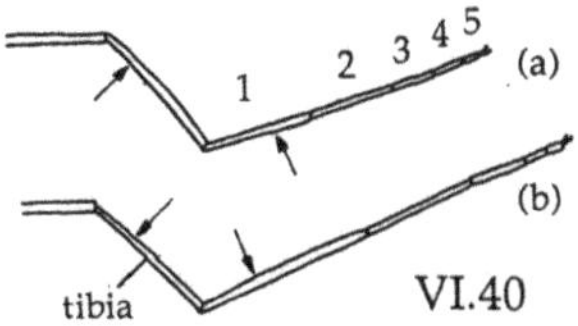

VI.40

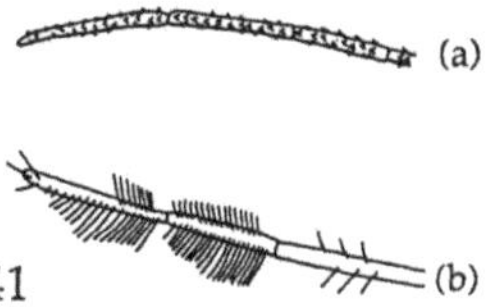

VI.41

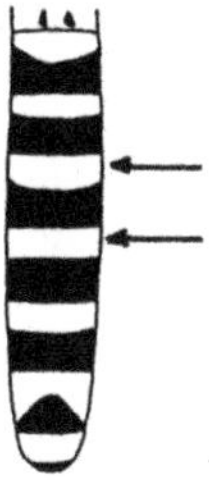

VI.42

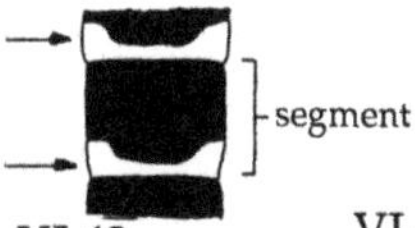

VI.43

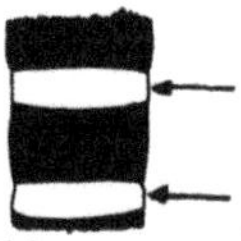

VI.44

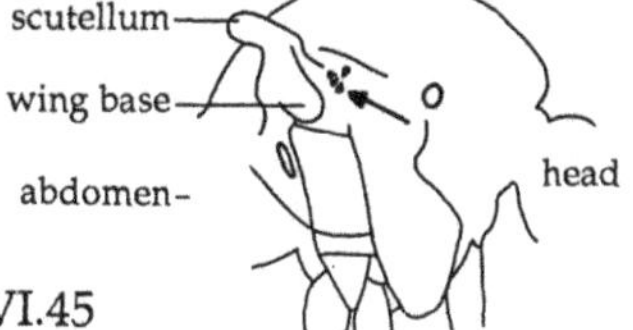

VI.45

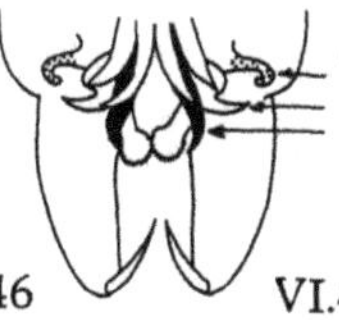

VI.46

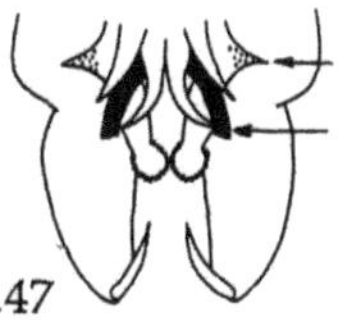

VI.47

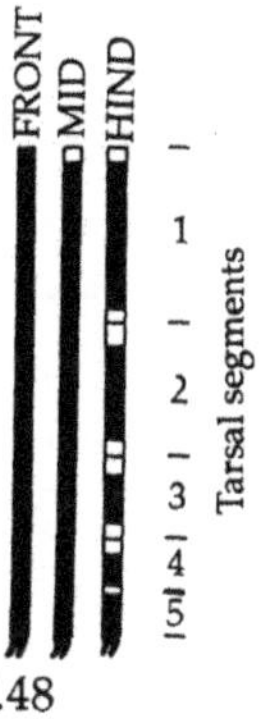

VI.48

VI.49

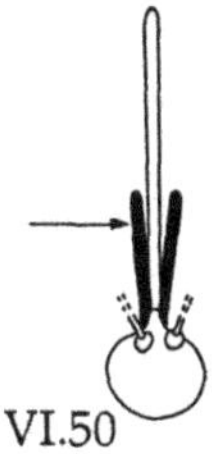

VI.50

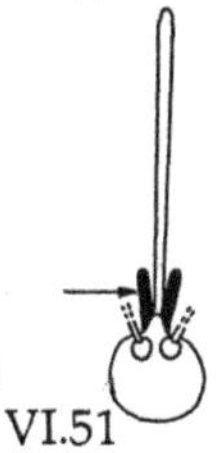

VI.51

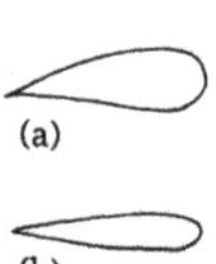

VI.52

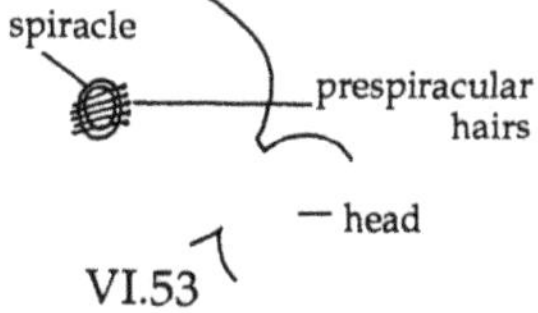

VI.53

22 Scutum with distinct stripes of white scales, resembling a lyre in shape (VI.54); femora and tibiae with white scales aggregated into conspicuous spots or stripes (VI.55a); front margin of costa (VI.56) almost entirely pale-scaled; male palps shorter than proboscis *Culiseta longiareolata* (p. 45)

– Scutum with pale scales arranged in a more diffuse pattern; femora and tibiae with dark and pale scales intermixed (VI.55b); front margin of costa (VI.56) mainly dark or with a mixture of dark and pale scales; male palps longer than proboscis 23

VI.54

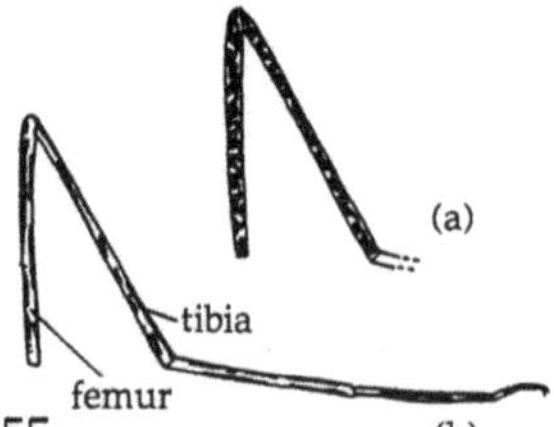

VI.55

23 Dark scales on wings aggregated to form dark spots (VI.57); rings of pale scales on tarsal segments broad (VI.58a); postspiracular area scaled (VI.59) subgenus *Culiseta* 24

– Wings without dark spots; rings of pale scales on tarsal segments narrow (VI.58b); postspiracular area without scales subgenus *Culicella* 26

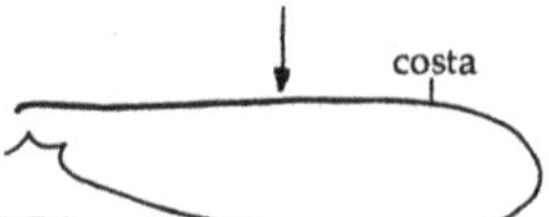

VI.56

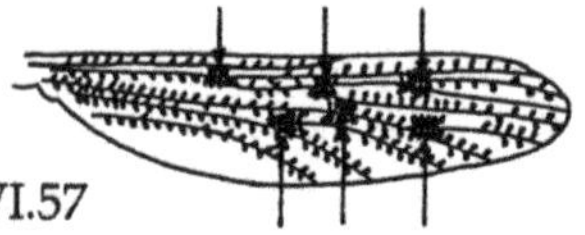
VI.57

24 First tarsal segments without central pale ring (VI.60a) *Culiseta alaskaensis* (p. 46)

– First tarsal segments with central pale ring (VI.60b) 25

25 Ground colour of abdomen above uniformly black, with white or yellowish bands *Culiseta annulata* (p. 46, pl. 2)

– Ground colour of abdomen above black, generally with scattered yellowish scales and with yellowish bands *Culiseta subochrea* (p. 46)

Note. It is very difficult to distinguish these species. The hypopygia of males appear to be identical.

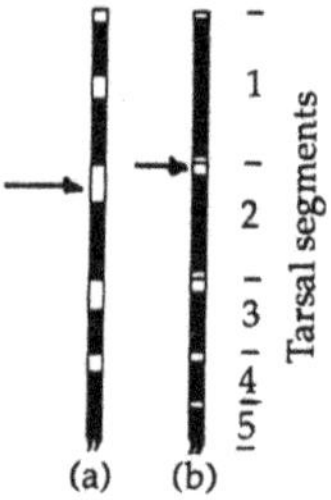

VI.58

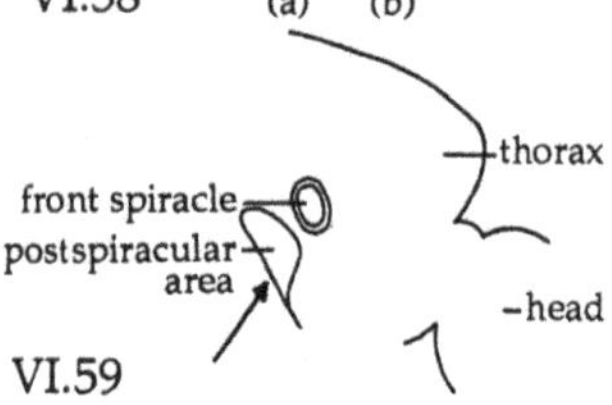

VI.59

26 Female proboscis dark with, at most, a few scattered pale scales (VI.61a); last two hind tarsal segments of female without pale rings (VI.62a); dark scales on lower surface of abdomen scattered, not forming patterns (VI.63) *Culiseta morsitans* (p. 45)

– Female proboscis dark but with conspicuous aggregations of pale scales (VI.61b); last two hind tarsal segments of female with or without pale rings (VI.62b); dark scales on lower surface of abdomen forming V-shaped patterns (VI.64) 27

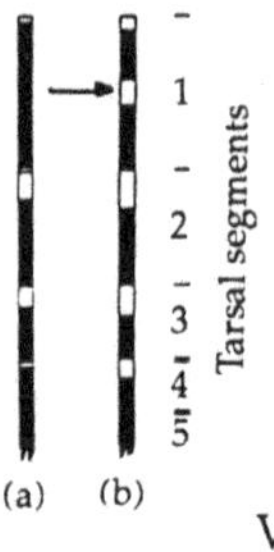

VI.60

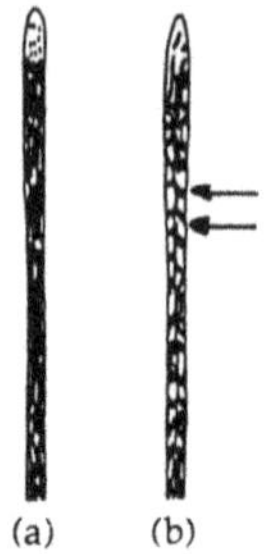

VI.61

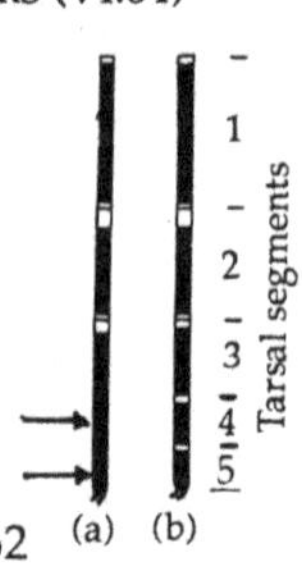

VI.62

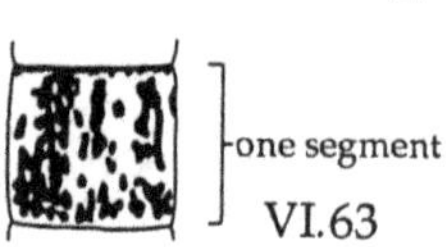

VI.63

VI.64

27 Pale rings normally present on all tarsal segments (VI.65a); male palps longer than proboscis by almost the whole length of the last segment (VI.66a); male hypopygium with basal lobe of gonocoxite with 3–5 spines stouter than the other hairs and none of them reaching to the tip of the gonocoxite (VI.67)
Culiseta fumipennis (p. 45)

– Pale rings absent or inconspicuous on last two segments of hind tarsi (VI.65b); male palps longer than proboscis by only about half the length of the last segment (VI.66b); male hypopygium with basal lobe of gonocoxite with two or three spines stouter than the other hairs and one of them reaching the tip of the gonocoxite (VI.68) *Culiseta litorea* (p.45)

28 Prominent tufts of white or whitish scales on the top of the head and on the front margin of the scutum; scutum darker at the sides (VI.69) 29

– Without such prominent white or whitish tufts, scales in these regions brown; scutum not obviously darker at the sides (VI.70) *Anopheles algeriensis* (p. 36)

29 Wings with aggregations of dark scales forming spots (VI.57) *Anopheles atroparvus*
and *Anopheles messeae* (p. 36, pl. 1)

Note. It is not possible to distinguish these two species as adults.

– Wings without spots, dark scales distributed evenly on veins 30

30 Pale scales on top of head and front margin of scutum (VI.71) pure white; general coloration of scutum dark greyish-black; male hypopygium with two parabasal spines, both unbranched; internal spine arising near the centre of the gonocoxite (VI.72)
Anopheles plumbeus (p. 37, pl. 1)

– Pale scales on top of head and front margin of scutum off-white; general coloration of scutum brown; male hypopygium with three parabasal spines, two of which are branched; internal spine arising near the tip of the gonocoxite (VI.73) *Anopheles claviger* (p. 37)

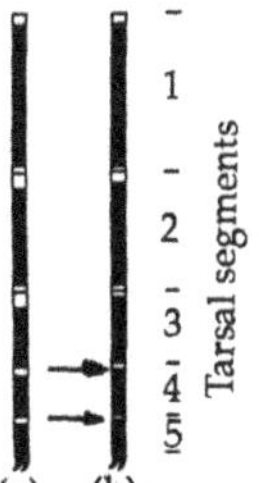

VI.65

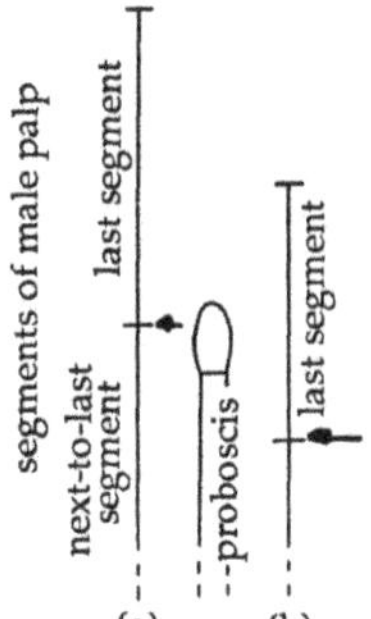

VI.66

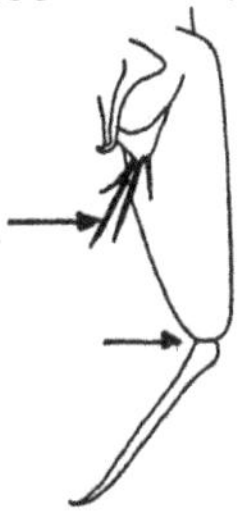
VI.67

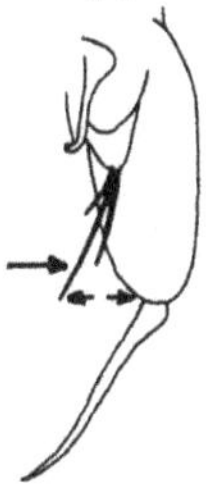
VI.68

VI.69

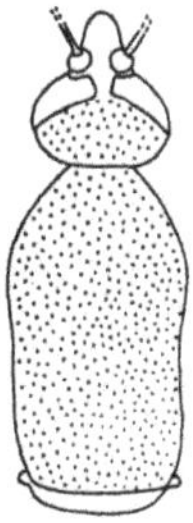
VI.70

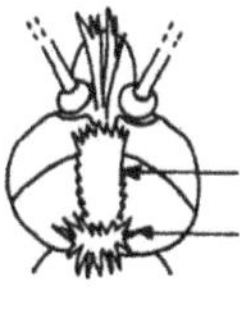
VI.71

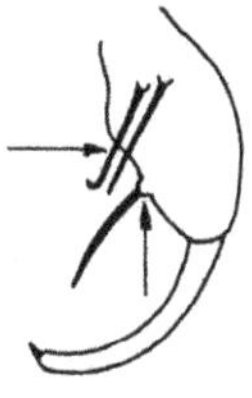
VI.72

VI.73

4 Notes on the British mosquitoes

In this section brief details are given of the ecology and, where appropriate, the important distinguishing features of the various life stages. This section should be used in conjunction with the keys to give a picture of the species that takes into account morphological and ecological features. The keys use structural characters almost exclusively and details of, say, preferred aquatic sites may be important in arriving at a final identification.

However, as so little is known about the ecology of British mosquitoes the details given here may well need considerable modification in the light of future investigations. So, for example, the fact that only one species of *Aedes* has ever been recorded from tree holes in this country does not mean that all *Aedes* recovered from tree holes should be assigned to *Aedes geniculatus* without question. Similarly, because a species has never been recorded from a given locality does not mean that it does not occur there; perhaps no one has ever looked.

These notes highlight aspects of the ecology of British mosquitoes where limited information exists and further studies are needed.

Anopheles algeriensis (p. 35)

Adults of this species are reddish-brown and easily separated from the other British *Anopheles* by the absence of wing spots and tufts of pale scales on the head and scutum. The most conspicuous larval characteristic, distinguishing it from other British *Anopheles*, is the dark head markings which form three distinct bands.

The main distribution of *An. algeriensis* is in the Mediterranean countries, but it was recorded in Norfolk as adults and larvae in shallow puddles at the margin of a marsh in 1932 and for a further 15–20 years. More recently, in 1987, it was again discovered at Cors Goch in Anglesey, larvae being found in shallow, unpolluted water among emergent sedges and rushes in fenland.

In other parts of its range larvae are found in both natural and artificial collections of fresh and brackish waters. In Europe it overwinters in the larval stage. Adults bite humans readily in the open, entering houses only rarely.

Anopheles atroparvus and *Anopheles messeae* (p. 35; pl. 1)

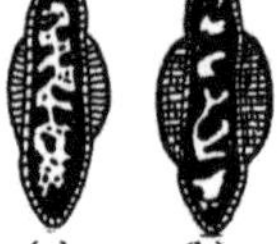

(a) (b)

Fig. 31. Pattern on deck (flattened surface) of egg of (a) *Anopheles atroparvus* and (b) *An. messeae*.

These two species are members of the *Anopheles maculipennis* complex and are morphologically identical except for the patterning of the eggs (fig. 31). The combination of dark wing spots and uniformly dark legs and abdomen differentiates these two species from all other

British mosquitoes. Later larvae are characterised by the design of the pecten and the features of certain hairs on the head.

Aquatic stages develop in relatively clean, permanent, standing or very slow-moving waters with surface or emergent vegetation. *An. messeae* is found in fresh water but *An. atroparvus* prefers brackish-water pools and ditches.

Both species rest in houses, feeding on the occupants, but whereas in the last century human dwellings were often heavily infested, modern houses are less attractive to these mosquitoes and they rest and feed mainly in animal shelters. Only females which have not laid eggs (nullipars) survive to overwinter. *An. messeae* seeks cold shelters, normally unoccupied by people, and undergoes complete hibernation, surviving on its food reserves. *An. atroparvus*, however, enters warm animal houses and sometimes human dwellings and feeds periodically.

Eggs are laid from April onwards and adults from these appear in May or June. There are two or three generations per year.

Because of the difficulty in separating these species, most older records are of little value. Both species appear to be found throughout Britain, perhaps less commonly in the north, *An. atroparvus* having a mainly coastal and estuarine distribution.

Anopheles claviger (p. 35)

Although similar in size and colour to *An. atroparvus* and *An. messeae*, this species is distinguished from them by the absence of wing spots. It shares with *An. plumbeus* the pale scales on the head and scutum (unlike *An. algeriensis* in which the scales are dark) but whereas *An. plumbeus* is a small, blackish mosquito, *An. claviger* is generally larger and brownish and the contrast between light and dark scales is less pronounced. Larvae of *An. claviger* are recognised by their head hairs and pecten characters.

An. claviger exploits a variety of fresh and brackish water habitats including ponds, ditches and rain barrels, usually in shaded situations. Overwintering is as larvae and the first adults of the year appear in March, or exceptionally February. There may be two or more generations per year. Adults may shelter indoors early in the year but rest outdoors as the weather becomes more clement. Females feed on various mammals and bite people readily, chiefly in the open.

An. claviger is widely distributed throughout Britain.

Anopheles plumbeus (p. 35; pl. 1)

The unspotted wings, pure white scales on the head and scutum and dark coloration differentiate this small mosquito from other *Anopheles*. Aquatic stages are found almost exclusively in tree holes and this species is common throughout Britain wherever there are mature trees with water-filled holes.

In common with the other two tree-hole mosquitoes (*Aedes geniculatus* and *Orthopodomyia pulcripalpis*), and in contrast to all other British species, the antennae of larvae do not have spines.

The adult season extends from April or early May until late autumn. Eggs are laid on the sides of tree holes just above the water line and do not hatch until flooded. There may be one or several generations per year. Winter is normally passed as larvae although the eggs are frost-resistant. Pupation occurs in the early spring.

An. plumbeus is a persistent biter, especially around dusk, both in and out of doors.

Aedes cinereus (p. 30)

Adults of *Ae. cinereus* are almost entirely blackish brown except for a reddish thorax and two narrow pale stripes on the sides of the abdomen. Larvae have extremely long anal papillae, a feature shared only with *Ae. sticticus*.

The one-year life cycle is typical of *Aedes* with oviposition occurring in the summer in dry areas prone to freshwater flooding. Larvae hatch in spring into flooded meadows, ponds, ditches and marshes and their development is normally completed by May with adults appearing by June and present until August or September.

Females feed on mammals including man, at dusk and during the night and daytime in shaded situations.

Ae. cinereus has a widespread, but patchy, distribution in England, Wales and Scotland. It is locally common in some areas where it may be a troublesome biter.

Aedes vexans (p. 32)

In general appearance adult *Ae. vexans* resemble *Ae. cantans* although there are several differences, the most obvious being that in *Ae. vexans* the pale abdominal bands narrow in the centre. Larvae are characterised by the comb which has very few scales and the pecten with some widely spaced spines.

Although a very abundant pest species throughout the world, it has been reported on only a few occasions in Britain from southern England and Wales. Little is known of its ecology and life cycle in Britain but elsewhere it has a typical *Aedes* life cycle breeding in temporary, unshaded collections of water. Adults bite persistently in the daytime and may be a source of serious annoyance to people and domestic animals.

Aedes geniculatus (p. 30; pl. 3)

This striking mosquito has conspicuous white knee spots, a white pattern on the thorax and white patches on the abdomen which do not form complete bands. Larvae, too, have many distinguishing characters, among them smooth antennae and evenly aligned comb scales.

The aquatic stages are found in water-filled tree holes, especially in beech. Eggs are laid on the sides of the cavities and await submergence after rainfall. *Ae. geniculatus* is most common in larger rot holes but is also found in smaller holes, often at the base of trunks. Overwintering may be as either eggs or larvae although only eggs can withstand freezing.

Adult emergence begins in May and there are often two generations per year with the last adults of the season occurring in September. Females bite people readily but seldom enter houses. Although common in most English woodlands, *Ae. geniculatus* has not been recorded from Ireland, Scotland or Wales.

Aedes annulipes (p. 32)

This large mosquito is similar in most respects to *Ae. cantans* from which it is distinguished with difficulty. These two species differ only in details of male genitalia and coloration, the black and white of *Ae. cantans* being replaced by dark brown and yellowish-white. At present there is no way of separating larvae and pupae of the two species.

Ae. annulipes has a 1-year life cycle and eggs are laid in depressions in open or partially shaded situations subject to flooding by autumn and winter rain. The eggs hatch from December onwards and larval development is normally completed by April or May, adults appearing shortly after. Females feed principally on mammals and bite people by day as well as in the evening.

Ae. annulipes has been recorded in Britain from England only. Because of its great similarity to *Ae. cantans* it may have been overlooked in some areas.

Aedes cantans (p. 32)

As indicated above, this species is morphologically very similar to *Ae. annulipes*. In its life cycle too the similarities continue except that the aquatic sites are normally shaded temporary freshwater pools.

Females are most active just after sunset but will bite various mammals including man in woodlands in the daytime. *Ae. cantans* does not normally enter houses. This species has been recorded in many areas of England and also in Wales and Scotland.

Aedes caspius (p. 31)

Ae. caspius is the most ornate of British mosquitoes with its mainly fawn scutum divided into thirds by two parallel white stripes, its broad pale abdominal bands and its pale leg rings which embrace both ends of the tarsal segments. Larvae have extremely short anal papillae in common with *Ae. detritus* and several other species.

Once again the egg is the overwintering stage and hatching occurs after flooding in the spring. The egg-laying sites are usually coastal marshes, although there are many records of *Ae. caspius* from inland, non-saline waters. The first adults often appear in April, within a month of hatching, and there may be several generations each year depending on the tidal and rainfall patterns. Adults may still be present in October. Females bite man viciously mainly out-of-doors but they sometimes enter houses to feed.

Ae. caspius is widespread in many coastal districts of Britain with, in addition, several inland records in England.

Aedes communis (p. 31)

Although widely distributed in many other parts of Europe, *Ae. communis* has been recorded only twice from Britain, once from Nottinghamshire and once from Jersey. Adults of this species closely resemble *Ae. detritus*, *Ae. sticticus* and *Ae. leucomelas* and larvae are similar to those of *Ae. annulipes* and *Ae. cantans*.

In continental Europe the life cycle resembles that of *Aedes punctor* and females bite people viciously in woodlands.

Aedes detritus (p. 31; pl. 3)

A small brown species, this has a banded abdomen and a peppering of white scales over the body, wings and legs but lacks pale leg rings. Larvae have extremely short anal papillae in common with several other species, including *Ae. caspius*, and a large number of comb scales.

Ae. detritus oviposits in salty ground prone to periodic flooding and usually a generation follows each immersion. Not all of the eggs hatch after the first immersion and some require several cycles of alternate flooding and drying before they will hatch. Some eggs laid in dry areas in the summer and autumn may hatch in the winter and so both eggs and larvae may overwinter. Eggs which are not submerged may survive for over a year and hatch when eventually flooded. The overwintering larvae develop to adults in March and there are successive generations throughout the summer and autumn.

Females bite man persistently chiefly out-of-doors but will also enter buildings to feed. In many coastal areas this salt-marsh species creates a greater nuisance than any other British mosquito.

Ae. detritus has a wide but patchy distribution in low-lying coastal areas and some inland saline waters of Britain and has been recorded from England, Wales and Scotland.

Aedes dorsalis (p. 31)

This species resembles *Ae. caspius* but differs in coloration in several respects. Thus the scutum is ashy-white with a dark brown central area and the abdomen is black and ashy-white. As in *Ae. caspius* the anal papillae of the larvae are extremely short.

The life cycle of *Ae. dorsalis* is essentially similar to that of *Ae. caspius* although the adult season is restricted from May to September. Larval sites are sunlit, or partially shaded, temporary collections of fresh or brackish water in ditches and pools. Females feed on a variety of mammals including man and bite during the daytime and evening in open areas.

In Britain *Ae. dorsalis* is known only from England as far north as Cumbria.

Aedes flavescens (p. 32)

Ae. flavescens has a golden-brown scutum, a yellow abdomen with dark lines in the centre and down the sides and broad pale bands on the legs. Larvae belong to the 'short anal papillae' group and have distinctively pointed comb scales.

Larval sites are always open and sunlit but may be fresh or brackish and include ditches, flooded meadows and marshes. Adult emergence begins in May and adults are rarely encountered after July. Females feed on larger mammals including man and may be troublesome locally.

Eggs are laid in areas subject to flooding. Whereas those immersed in autumn or winter do not hatch until February or March, eggs laid in June will hatch in summer if immersed. Thus this species, although usually having a 1-year life cycle, may produce two generations a year.

Ae. flavescens has been recorded in coastal Essex and Kent and at Barton on Humber. In the Isle of Grain *Ae. flavescens* has become increasingly common in recent years while *Ae. caspius*, once abundant, is now rare.

Aedes leucomelas (p. 31)

Adults of this rare species resemble *Ae. detritus* closely but lack the scattered pale scales on the abdomen. The only authenticated record of this species in Britain is from Nottinghamshire in 1919.

In continental Europe larvae are found in unshaded or partially shaded fresh and brackish water both on coasts and inland. The life cycle resembles that of *Ae. caspius*.

Aedes punctor (p.31)

This small, dark-legged woodland mosquito has pale abdominal bands which characteristically narrow in the centre. The mature larvae differ from all other British *Aedes* as the saddle completely surrounds the anal segments. Eggs are laid in dried-up hollows and ditches or above the level of standing water during the summer months. They are flooded in autumn and winter and hatch from December onwards, the first adults emerging in April or May. Only one generation occurs each year and the last adults of the season are seen in late August or September.

Females feed on a range of mammals including humans and bite readily, especially at dusk in the neighbourhood of aquatic sites. There are some records of *Ae. punctor* biting in houses.

Ae. punctor is widely distributed in Britain dependent on the availability of temporary woodland pools.

Aedes rusticus (p. 30)

This species has a very similar life cycle to that of *Ae. punctor* and *Ae. cantans* and often shares similar woodland aquatic sites although those selected by *Ae. rusticus* seem always to be lined with dead leaves. Like these other two species, *Ae. rusticus* is a troublesome biter during the summer in the open near its aquatic sites. However, *Ae. rusticus* appears to have a shorter adult season, being seen only rarely after July.

Ae. rusticus is widely distributed in England with a few records from Scotland and Wales. Like *Ae. punctor* it is a dark-legged species with a banded abdomen. However, the abdominal pale bands widen centrally rather than narrowing as they do in *Ae. punctor.* In the later larval stages there are 3–4 pairs of hairs on the dorsal surface of the siphon and the pecten overlaps the siphonal hair, both characters being diagnostic.

Aedes sticticus (p. 30)

There are no recent reports of this species, past records being from only four widely separated areas ranging from the New Forest to Scotland. *Ae. sticticus* is dark-legged with parallel pale bands on the abdomen and a white scutum with a broad dark brown central stripe. Larvae, like those of *Ae. cinereus,* have extremely long anal papillae.

From observations elsewhere in the world it appears that *Ae. sticticus* overwinters in the egg stage and that the larvae develop in open and partly shaded temporary waters. Females bite people readily in open situations.

Coquillettidia richiardii (p. 33; pl. 4)

Adults of this species are reddish-brown with striped legs; the pale abdominal bands, when present, are inconspicuous. It shares a characteristic pale ring in the middle of the first tarsal segments with *Cs. annulata* and *Cs. subochrea* but, unlike these two species, *Cq. richiardii* does not have dark spots on its wings.

Larvae and pupae are unique among the British mosquitoes in possessing siphons and trumpets designed to pierce roots and stems of water plants, from which they obtain their oxygen. Eggs are laid in rafts on the water surface from May to August and the young larvae descend to attach to sweet flag (*Acorus*), water crowfoot (*Ranunculus*), the grass *Glyceria* or reed mace (*Typha*) as soon as they emerge. Although some larvae reach the fourth instar by late summer, pupation does not occur until the following spring. Pupae remain attached to plants, ascending only to emerge. Adults appear in May and have been recorded as late as September, females feeding mainly on mammals and biting humans inside and outside buildings. There is only one generation per year.

Cq. richiardii is locally common all over Britain and often abundant where it occurs.

Culex modestus (p. 33)

The only records of this species in Britain are from Portsmouth, Gosport and Hayling Island in 1944–45 where it was found in fresh and brackish water sites. Almost certainly it was introduced from its native Mediterranean area. Little is known of its ecology and biology in Britain but elsewhere in its range it bites man viciously.

Adults may be distinguished from other *Culex* species by the absence of pale abdominal bands, while in larvae the ventral siphonal hairs are arranged in a line and bear many branches.

Culex pipiens (p. 33; pl. 2)

'typical' form

This small mosquito is common throughout Britain and adults are often seen resting in houses. It has dark legs and distinct white bands on the abdomen and is morphologically very similar to *Cx. torrentium*. Egg rafts are laid in a variety of habitats including ponds, marshes and water butts. They first appear in early April and larvae develop to give adults from May onwards. These adults feed almost exclusively on birds with some records of feeding on frogs, lizards and snakes. They produce further egg rafts through to September but this 'summer generation' does not survive beyond late autumn.

Females emerging after August feed exclusively on plant juices and lay down fat bodies. Although fertilised, they do not feed on blood and their ovaries do not develop until the following spring. Instead they seek suitable hibernation sites which are cool, damp places such as cellars and outbuildings where they remain until the following spring. Emergence from hibernation occurs in spring and egg rafts are soon laid. Overwintered females seem to have a rather short life expectancy after reactivation in the spring, with very few surviving beyond late May.

***'molestus'* form**

Until recently this was considered to be a distinct species and was called *Culex molestus*, but it is now treated as a form of *Cx. pipiens*. It is identical to the typical form morphologically in all details but differs biologically in five important respects: it is man-biting; the aquatic stages are usually found underground, for example in flooded cellars and boiler rooms; it can lay an egg batch without having had a previous blood meal (autogenous); it can mate in confined spaces such as in buildings (stenogamous); both larvae and adults remain active throughout winter and this form does not undergo diapause (a period of suspended development and growth).

This form of *Cx. pipiens* has been reported from many areas of England but it may well be more widely distributed in Britain generally.

Culex torrentium (p. 33)

This species is similar to *Cx. pipiens* in almost every respect and is separable with certainty only in the male, by examination of the hypopygium.

It often occupies the same larval sites as *Cx. pipiens* but seems better adapted to container-water. It has also been recorded on occasions from tree holes.

Adults do not appear to enter houses, and they feed exclusively on bird blood. Although adult females are the overwintering stage, as in *Cx. pipiens*, the hibernation sites of *Cx. torrentium* are unknown and are not those of *Cx. pipiens*. *Cx. torrentium* is ubiquitous in rural and peri-urban areas in southern England and has been reported as far north as Cumbria. There are, at present, no records from Ireland, Scotland or Wales.

Culex territans (p. 33)

Although British records of *Cx. territans* are scarce, they are widespread, ranging from the south coast of England to Scotland.

Egg rafts are produced as in other *Culex* species, but are laid above the water line, the larvae entering the water on hatching. Very little is known of the ecology of *Cx. territans* in Britain but the main aquatic habitat is probably small, permanent collections of water although other accumulations of water, both natural and artificial, may be used.

A few records exist of *Cx. territans* feeding on man, other mammals and birds but most are of feeding on amphibians and reptiles. Adult *Cx. territans* may be distinguished from all other British *Culex* as the pale abdominal bands are on the hind border of the segments. The siphons of *Cx. territans* larvae are characteristically very narrow.

Culiseta longiareolata (p. 34)

Cs. longiareolata adults have a characteristic reddish-brown scutum with white markings while the short siphon and arrangement of the pecten spines help identify the larvae.

Cs. longiareolata appears to be a very rare mosquito in this country and it is doubtful whether it is established here. It has been recorded on only three occasions, all from the south of England. In the Mediterranean area it develops in almost any type of water whether fresh or brackish, clean or polluted, artificial or natural. Little is known of the ecology or behaviour of *Cs. longiareolata* in Britain but it does not appear to bite humans.

Culiseta fumipennis (p. 35)

The appearance of adults of this species is similar to that of *Cs. litorea* and *Cs. morsitans* but larvae can be distinguished by the isolated spines of the pecten.

On limited data it would appear that the life cycle too is similar although larval records of *Cs. fumipennis* refer solely to unshaded freshwater bodies, both temporary and permanent, suggesting that it may be more restricted in its habitats than *Cs. morsitans*.

Cs. fumipennis females do not appear to bite humans or domestic animals and the species has been recorded widely in Britain.

Culiseta litorea (p. 35)

Cs. litorea adults closely resemble those of *Cs. morsitans* and *Cs. fumipennis* and larvae are difficult to distinguish from those of *Cs. morsitans*.

The life cycle is essentially the same as that of *Cs. morsitans* and it too can tolerate slightly brackish water. However, it has been found only in open waters.

Females feed mainly on birds, although they occasionally bite reptiles and mammals including man. Records of *Cs. litorea* in Britain are from southern England and Ireland.

Culiseta morsitans (p. 34)

Morphologically adult *Cs. morsitans* are almost identical to *Cs. fumipennis*. While the larvae of these two species can be distinguished by the spacing of the pecten spines, there is difficulty separating larvae of *Cs. morsitans* from those of *Cs. litorea*.

The life cycle of *Cs. morsitans* is similar to many *Aedes* species in that eggs are deposited during the summer months in dry hollows or above the level of standing water. The eggs hatch following immersion by autumn and winter rainfall. Larvae may hatch as early as November but pupation does not occur until spring. Larvae cannot withstand freezing, although they can survive under ice for long periods. There is only one generation per year.

Aquatic sites selected may be fresh or slightly brackish waters in ponds, ditches and pools in either open or shaded situations.

Adults have been recorded from May to October and the majority of females feed on birds with just a small percentage feeding on reptiles and mammals including man. *Cs. morsitans* does not appear to enter houses to feed. *Cs. morsitans* is widely distributed in England and Wales and has been recorded from Scotland and Ireland.

Culiseta alaskaensis (p. 34)

There have been only occasional reports of this species in Britain, all as adults from Scotland and England north of the Humber between the months of March and October. No ecological data exist for *Cs. alaskaensis* in Britain but elsewhere in its range larvae are found in open pools and adults can pose a serious human biting problem.

In appearance, adults closely resemble *Cs. annulata* but lack the pale rings near the tip of each femur and on the middle of the first tarsal segments. Larvae are similar to those of *Cs. annulata* and *Cs. subochrea*.

Culiseta annulata (p. 34; pl. 2)

Cs. annulata is both widespread and common throughout Britain which reflects its varied choice of aquatic sites. Egg rafts are laid in natural or artificial waters in sunlit or shaded areas. The water may be fresh or brackish and clean or, more often, polluted. Examples of such sites are ponds, ditches, drains, marshes, garden water butts and cisterns. There may be several generations each year.

Adults are often found in houses where they feed readily on the occupants. They also bite out-of-doors, attacking man, other mammals and birds.

Cs. annulata is able to overwinter in any stage under suitable conditions. Females do not enter diapause but vary between quiescence and cycles of feeding and egg-laying. Experiences of mosquito bites between late autumn and early spring are almost always attributable to *Cs. annulata*.

This is among the largest of the British mosquitoes and is easily recognised by its size, spotted wings and white rings on the legs. Larvae have a basal siphonal hair and hair-like spines in the pecten.

Culiseta subochrea (p. 34)

This mosquito is uncommon in Britain having been recorded from England only. It resembles *Cs. annulata* closely in form in all stages apart from the more yellowish adult coloration, details of the male genitalia and the fringing of the pupal paddles (fig. 32). There are no apparent differences in the aquatic habitats and life history of this species and *Cs. annulata* except that egg-laying without a prior bloodmeal (autogeny) has exceptionally been observed in *Cs. subochrea* but not in *Cs. annulata*.

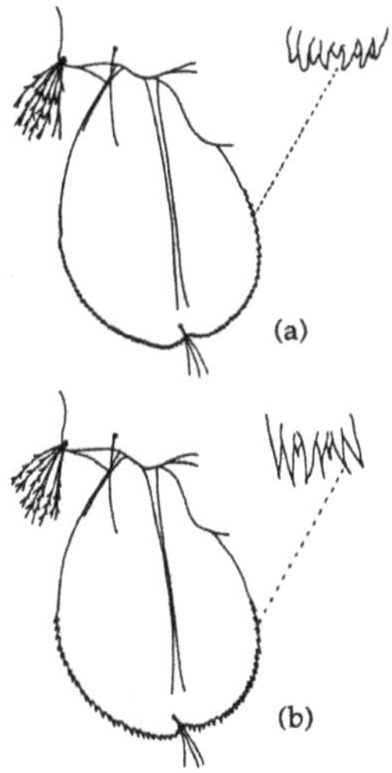
Fig. 32. Shape of spines on edges of pupal paddle in (a) *Culiseta annulata* and (b) *Cs. subochrea*.

Orthopodomyia pulcripalpis (p. 33; pl. 4)

The adult mosquito is easily recognised by the black and white patterning of the scutum, abdomen and hind legs and the relatively long female palps. Larvae, too, have their peculiarities: this is the only British species to lack a pecten apart from *Cq. richiardii* in which the siphon is highly modified.

Or. pulcripalpis breeds in water-filled tree holes, especially those in beech, elm and horse chestnut. North American species of *Orthopodomyia* lay their eggs singly on the sides of tree holes close to the water line and the larvae descend into the water on hatching, but whether this is true of *Or. pulcripalpis* is not known.

In Britain larvae are the main overwintering stage and these develop to pupae by June–July and adults emerge soon after. Only one generation occurs each year. Little is known of adult ecology and behaviour but females appear to feed on birds.

The few distribution records which exist for this species are all from the south of England.

5 Studying mosquitoes

Collecting and sampling mosquitoes

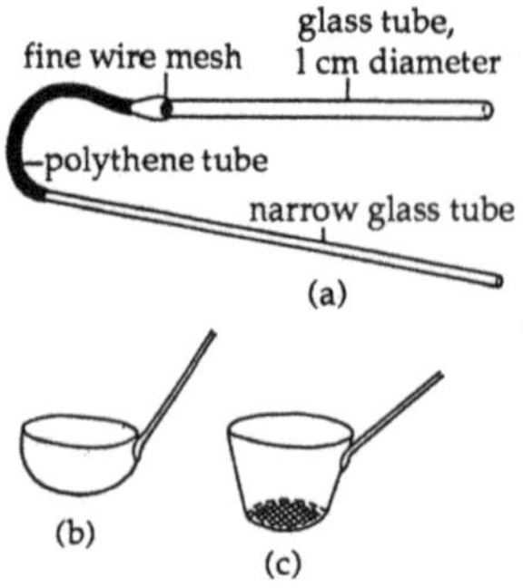

Fig. 33. Collecting equipment. (a) Aspirator or 'pooter' for collecting adults; (b) ladle; (c) modified saucepan with mesh bottom. (b) and (c) are for collecting larvae and pupae.

Many different methods have been devised for sampling both the adults and aquatic stages of mosquitoes. Most of these are described by Service (1976). It is difficult to assess mosquito densities precisely but estimates may be made using quite simple techniques employing easily obtainable equipment (fig. 33). The dipper, a kitchen ladle, is often the basis for both collecting and estimating the densities of larvae and pupae. This is best used for surface-feeding species, such as *Anopheles*, rather than browsers, like *Aedes*, which spend much of the time submerged. Netting deeper into the water with a stout net for a prescribed time can be used for comparative studies and appears to provide reliable estimates of relative population density. Species occurring in tree holes and other inaccessible places may be collected by sucking out the water with a suitable aspirator. Eggs of *Aedes* and *Culiseta* (*Culicella*) may be extracted from soil and leaf litter from known egg-laying sites in late summer and autumn before they are submerged. This is done by collecting samples of soil or leaf litter lining these sites and washing them through graded sieves to remove particles larger than a few millimetres and then searching through the remaining material with the aid of a hand lens or dissecting microscope.

Although routine dipping can give an indication of larval densities, it may be desirable to gain a more detailed estimate. In a study of the densities of *Aedes communis* in North America, Fish & Joslyn (1984) were able to correlate larval density with surface area of woodland snow pools. A similar study could be tried with our woodland *Aedes* to see whether the correlation holds for these species. The study method was established by Joslyn, Conrad & Slavin (1985) in experiments with *Aedes sollicitans*. It has long been known that mosquito larvae immersed in certain vital stains, for example methylene blue, incorporate these into their tissues. Preliminary trials must be carried out to find the optimal concentration for a species, producing permanent blue marking of larvae, pupae and emergent adults whilst having little adverse effect on their development. For *Aedes punctor* a concentration of 100 mg per litre (= 100 parts per million (ppm) of methylene blue) would appear to be suitable.

The stain is visible to the naked eye, the larvae and pupae appearing distinctly blue. However, adults must be squashed on a white background and examined at x10 to x20 magnification. The colour appears to remain in the adult tissues for the duration of their life. The retention of the dye is not affected by sugar-feeding, blood-feeding or egg-laying.

In the application of the technique to field conditions Fish & Joslyn (1984) removed larvae of *Ae. communis* from a site and exposed them to a high concentration of Giemsa stain (100 ppm) for 1–3 hours. After washing, the larvae were returned to the pool, and thoroughly mixed with the population by agitation of the water while replacing the larvae. Approximately 18 hours later samples of larvae were removed and examined for blue colour. A modified Lincoln Index (see Chalmers & Parker, 1986) was then used to calculate the total population of the pool (*N*):

$$N = r\left[\frac{n+1}{m+1}\right]$$

where r = number initially marked; n = sample size; m = number of marked larvae in sample.

When releasing marked larvae it is imperative that there is complete mingling with the population. This may be best achieved by introducing the marked larvae evenly over the surface of the water after dividing the pool into a grid. Recapture should be made as soon as intermixing is believed to have taken place and before predation and other sources of population loss have significantly affected the introduced larvae.

For the study of some tropical mosquitoes, radioisotope labelling has been used and recaptured larvae identified by their effect on photographic film. Obviously this technique cannot be used without sophisticated equipment and official permission.

It would be worthwhile comparing the estimates of larval density obtained by the staining technique with those obtained by other methods. One alternative procedure is to enclose an area of pool with a frame such as a large bucket or oil drum which has had the bottom removed. This is pushed firmly into the bottom mud and all of the larvae collected from the delimited area by dipping. Several such counts need to be made to allow for sampling errors and to assess the patchiness of the population. Details of overall distribution, density of population, intermixing of species and instars and positional preferences of individual species can be found by such studies.

If the larval habitat is small, for instance a rain barrel or a tree hole, then it is feasible to remove all of the water, including any bottom sediment, and pass it through a sieve, or a series of sieves, to collect all of the larvae and pupae. Clean water is afterwards added to the site and this too is strained to collect any remaining insects. By this method absolute counts can be made for comparison with estimates based on samples. If the site is to be studied over a period of time, the larvae and the natural water can be returned after counting and examination and the numbers and instars reassessed at intervals. A technique similar to this, although not employing removal of the total volume of water, was used by Fallis & Snow (1983c) to follow the larval development of *Orthopodomyia pulcripalpis* (then written *pulchripalpis)* over a 2-year period in a tree hole in a

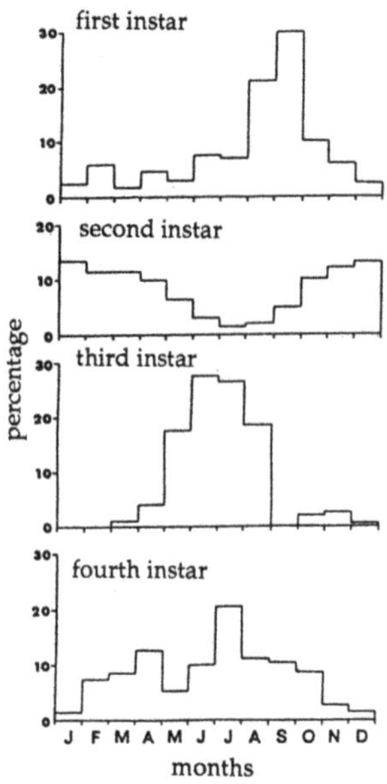

Fig. 34. Seasonal variation in the proportion of first- to fourth-instar larvae of *Orthopodomyia pulcripalpis* in a tree hole during 1979–81.

woodland. This study revealed some interesting data and showed the times of appearance of the larval instars and pupa and the onset of egg-laying. The study showed that the second stage larva is the main overwintering form and that the species has a single annual generation (fig. 34).

Recently there has been interest in aquatic light traps which consist of plastic tubes with funnelled entrances containing a light source, conveniently a chemical light stick. Quite how efficient these are and how well they assess populations is not known. Neither is it known whether all species are attracted or whether all instars of a given species are attracted equally. Some preliminary work suggests that there are distinct differences between species in the responses to light and that such traps would give biased samples, but this requires much more investigation.

Trap design and evaluation is an area where the interested collector, with a moderate ability to improvise, could make valuable contributions. The inventive collector could not only improve upon existing trap designs but also propose new methods of capture and assessment. As an example of inventiveness, the use of underwater adhesives has been looked into for holding larvae and pupae attracted to light sources (Service, 1984).

For adults the traditional method of assessing population numbers is to count the females landing on a human to feed, and express this as a number per unit time. For thoroughness this catch should be performed over a 24-hour period, although 1 hour before sunset will give useful data. This method, of course, works only for those species which feed on people and for those females which are at a precise physiological stage. Alternatively, adults may be caught in animal-baited traps, in malaise traps for sampling flying adults, or by collecting from known resting sites, such as vegetation for woodland *Aedes* or inside outbuildings for certain *Anopheles* and *Culex*. Light traps do not appear to attract many British species of mosquito but these are improved when combined with a carbon dioxide source. Again this aspect of trapping has received limited attention in Britain.

Mark-release-recapture experiments may also be performed on adult mosquitoes to assess densities, dispersal, longevity and other biological details. For example, females marked on emergence could be recaptured at their first attempt to bite in the early adult season and hence the time elapsing between emergence and the first blood-meal assessed. Also males from a swarm could be captured by netting, marked and released at the same site to see whether the same individuals constituted the swarm on subsequent days.

Two methods are readily available for marking adults, the use of radioisotopes being an impractical third. Marking with paints is widely used but very time-consuming. However, it has the great advantage that unique marks can be applied with a combination of colours and positions so allowing individuals to be recognised on recapture. The usual area marked is the thorax and the paints employed are water-miscible artists' powder paints. Instead of a brush, a fine wire microloop is used to apply the paint to mosquitoes lightly anaesthetised with either carbon dioxide or ether. This is best carried out with the help of a hand lens or dissecting microscope. When the loop is applied the insect normally remains stuck to the loop, necessitating a tap on the hand holding the loop to free it.

Recaptured insects are examined with a hand lens. It is advisable to retain a few of the originally marked mosquitoes to refer to when examining recaptured specimens. Although the choice of paint colour is somewhat arbitrary, it is advisable to select a bright colour which contrasts well with the scale colours. A preliminary trial will soon demonstrate the best colour and loop size to use.

For marking large numbers of mosquitoes coloured or fluorescent dusts are advised. Of these, fluorescent dusts have the advantage that they can be detected at a distance of several metres when illuminated with an ultraviolet lamp. This may be especially valuable in surveying and studying mosquitoes at night, and it appears that light of this wavelength does not modify their behaviour. This technique is suitable for long-term field investigations as the insects do not have to be collected in order to be observed.

Fluorescent dusts are available from chemical suppliers in several colours so, as an example, males and females could be coloured differently and any differences in their night-time resting sites discovered. It is also possible to view marked male swarms at dusk and into the darkness.

Dusting is achieved using a simple puffer. A small amount of the dust is placed in the bottom of the container and is blown onto mosquitoes in a mesh cage covered with a plastic bag. The dust adheres to the mosquitoes, especially between the leg bases, and makes the mosquitoes fluoresce brightly under ultraviolet (UV) light. Portable UV lights are readily obtainable from specialist pet fish shops.

Care must be taken when recapturing marked mosquitoes to prevent unmarked specimens from being contaminated with dust from marked insects. Recaptured mosquitoes should therefore be collected and stored individually and not allowed to come into contact with others.

Although the British Mosquito Recording Scheme was created in 1980 to promote recording and collate species distribution records, the lack of collectors in many parts of the country has meant that the scheme is developing slowly and the records are patchy. What is needed is for more collections to be made and local species lists produced. These can also be the basis for updating and adding new species to the record lists as they occur. New collectors and recorders will almost certainly require either help with their identifications or confirmation of identifications made. Such assistance can be provided by the British Mosquito Group (address on p. 63). Apart from species lists, other details of the collections are extremely useful, such as the characteristics of the aquatic site for larval and pupal records, perhaps with detailed descriptions and data such as pH. Other information which is easily obtained includes records of which species occur together, the densities of the larvae and pupae and whether they favour certain areas of the water mass. If they are aggregated then it may be possible to ascertain why.

Recording cards are issued by the Biological Records Centre (BRC) and distributed by the British Mosquito Recording Scheme. Such cards are eventually returned to the BRC and the details extracted to give up-to-date information on the ecology and distribution of mosquitoes in this country.

For those without the time or inclination for prolonged, detailed recording, even casual collecting and identification is valuable. All of these records, or even specimens if identification is not attempted or possible, can be forwarded to the national Mosquito Recording Scheme and eventually the aim of producing species distribution maps for Britain can be achieved. Recorders may like to form local recording schemes, perhaps as part of a local natural history group. Such schemes are the backbone of national programmes.

Examining and preserving mosquitoes

Eggs

Egg characteristics are not at present used in the identification of British mosquitoes except to differentiate the two members of the *An. maculipennis* complex, *An. atroparvus* and *An. messeae*.

After collection, eggs may be preserved in either MacGregor's solution (10 ml 5% borax; 10 ml 40% formaldehyde; 0.25 ml glycerine; 80 ml distilled water), 5% formaldehyde solution or 70% ethyl alcohol.

They may then be examined in the medium with a dissecting microscope. Alternatively, permanent preparations may be made by either (i) dehydrating in graded alcohols, clearing in Histoclear or a similar agent (approximately 15 minutes each step) and mounting in Histomount, Euparal or Canada Balsam or (ii) partially dehydrating in 70% alcohol (15 minutes) and transferring to polyvinyl lactophenol which serves as both a clearing and mounting medium. The tube containing the specimen in polyvinyl lactophenol must be kept in the dark, otherwise the medium will discolour. Once the specimen has cleared (approximately 1 hour) it is mounted and the medium allowed to dry, again in the dark. When the mountant has hardened it may be freely exposed to light.

Larvae and pupae

Larvae and pupae may be killed by immersion into hot water and preserved in MacGregor's solution, 5% formaldehyde or 70% alcohol. They may then be examined in the same medium at about x40 magnification using a dissecting microscope.

Permanent preparations of larvae are made using the method described for eggs except that, in the case of culicine larvae, it is necessary to cut off the last two abdominal segments when the specimen is in the mountant and arrange this section alongside the rest of the body. This permits the characteristics of the hind segments to be viewed more clearly (fig. 35).

An alternative, and preferred, method is to rear larvae until they moult and make a permanent preparation of the cast skin (exuviae) as detailed above. In this way a series of slides can be prepared showing the features of all the larval instars and of the pupa of a given individual. The adult which finally emerges may be pinned, as detailed below, to complete the collection.

Pupae reveal few of their taxonomic features when mounted whole and should always be studied from the cast skins. In order to show its features fully the exuviae is separated into cephalothorax (head and thorax) and abdomen (fig. 7, p. 5). The cephalothorax is split along the back at the rupture line through which the adult emerged and can be laid flat and mounted alongside the abdomen, with the outer surface uppermost (fig. 35).

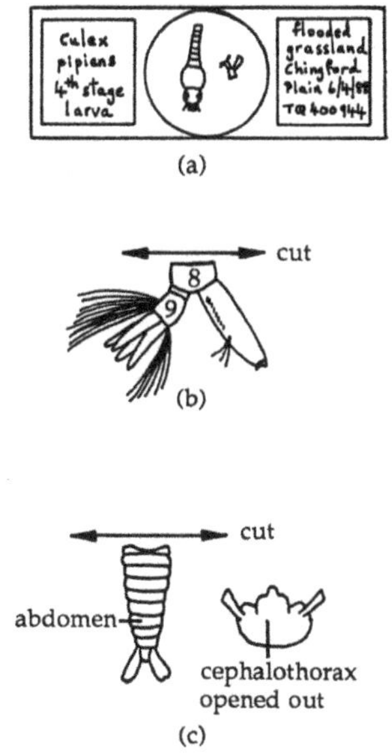

Fig. 35. (a) Prepared slide of culicine larva showing arrangement; (b) tip of abdomen of larva showing position of cut; (c) arrangement of pupal skin on a slide.

Adults

For taxonomic study the mosquito adult must be killed, preferably using ethyl acetate vapour. This is normally done by holding a small piece of cotton wool, which has absorbed a very small quantity of the chemical, over the open end of a glass tube containing the captured mosquito. Do not use a plastic tube as it will be attacked by the ethyl acetate. The mosquito is then pinned, preferably by the double-mounting method, and preserved dry. Double-mounting (fig. 36) entails the use of two pins for each specimen: (i) a small stainless steel point ('minuten') about 12 mm long, on which the specimen is mounted and attached to the stage; and (ii) a larger stainless steel pin, about 35 mm long, no. 3 in thickness, used to hold the stage and labels.

A number of materials may be used to form the stage, including thick card, cork strips or acetate sheet but the most suitable are polyporus (a white bracket fungus) or expanded polyethylene foam. These are cut into narrow strips about 2 mm x 2 mm and 13 mm long.

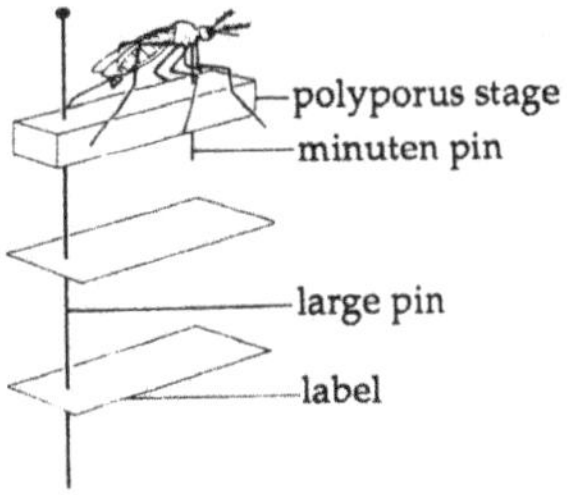

Fig. 36. Adult mounted using the 'double-pinning' method.

A 'minuten' pin is pushed almost completely through the strip about 2 mm from one end. The mosquito is placed on a sheet of white paper and, with the blunt end of the pin held with pinning forceps, the pin is inserted into the thorax of the mosquito from the ventral side midway between the legs. The pin is inserted until the point *just* emerges from the scutum. No attempt should be made to arrange the legs of the mosquito as they will almost certainly become damaged with loss of value of the specimen.

With the mosquito pinned, the polyporus or polyethylene foam strip is held carefully with forceps. The large pin is inserted from above about 1 mm from the end of the stage until about 10 mm of it remains above the stage. A pinning block is useful when performing this step and results in a uniform and neatly presented collection.

Labels are then affixed to the large pin (fig. 36, p. 53). These should be of good quality card and the details (including locality, date, collector's initials or name, and identification) written in Indian ink using a fine pen.

The pinned insect should be allowed to dry for 24 hours, in a desiccator if available, before being transferred to a storage box. The best storage boxes are made of wood and lined with cork. A cheaper alternative can be constructed by putting a tight-fitting floor of polyethylene foam into a stout cardboard or metal biscuit box. A small quantity of naphthalene should be added in order to prevent attack of the dried insects by pests. Preserved insects should be kept in the dark as prolonged exposure to sunlight may cause fading.

In order to identify male mosquitoes it is sometimes necessary to examine their external genitalia in detail using high power magnification. This entails separating the last few abdominal segments from the male and making a permanent slide preparation of this by clearing in warm 10% potassium hydroxide for about 15 minutes, neutralising in dilute acetic acid for 5 minutes and transferring to 70% alcohol. The specimen is then further processed as described above and eventually mounted, ventral surface uppermost (see p. 52), in either polyvinyl lactophenol or Histomount. The remainder of the male is then pinned in the usual manner. Both slide and pinned insect must be labelled to indicate their relationship.

Packaging and transport of mosquito specimens

At times it is necessary to send preserved specimens to others to share collections, for identification and for research. Whether sent by post or delivered by hand it is necessary to package them correctly in order to minimise any damage which could well make the specimens unidentifiable and of little subsequent value. Adults must always be transported dry and *not* in alcohol or other fluid preservative. Freshly killed adult mosquitoes should be packed on soft tissue paper or cellulose wadding in a small cardboard container such as a matchbox as follows: cut a layer of wadding or fold tissue to the size of the box and place a layer on the bottom of the box. Place the mosquitoes on this layer, taking care that they are well separated, and cover with another similar layer. This, in turn, may have specimens placed upon it and so on. The box is filled in this way. Take care not to mix specimens of different origin but either keep one box for a single collection or place paper labels on each layer if the box contains more than one collection. Data should also be included in or on the box as follows: date collected, name of collector, habitat, location (including Ordnance Survey map reference) and, if applicable, circumstances of collection. Here is an example:

23/6/88 C.D. Smith
deciduous woodland
HIGH BEACH, ESSEX.
51 510985
Biting human 9.45 p.m.

Packed in this way adult mosquitoes are well supported and should not become damaged. If the small box is then placed inside a larger box, surrounded by crumpled newspaper or chips of expanded polystyrene, this makes a package that can survive the postal system.

Larvae and pupae must be preserved in alcohol (about 70%) and packed in well-stoppered tubes, fully-filled to minimize damage to the specimens. A pencil-written paper label should be placed in the tube and this should be accompanied by another label, both giving full details. The tube is then packed in a cardboard box and completely surrounded by crumpled newspaper or expanded polystyrene chips. If several tubes are packed in the box they must be well separated. Once again label fully:

22/8/88 A.B. Jones
urban garden – rain water butt
CHINGFORD, LONDON, E.4.
51 500944

Rearing mosquitoes

Once collected, eggs, larvae and pupae can be reared on. This is best achieved by keeping them in water from the original site in a small container such as a jam jar; normally it is not necessary to add food. The jar may be maintained outside but preferably not in direct sunlight. However, to speed up the rate of development it may be kept indoors. If it is necessary to transfer the mosquitoes to clean water then either distilled water or rain water is preferred. If tap water is used then it should be allowed to stand for a few days before being used and food in the form of liver powder or dried yeast added in very small quantities. Liver tablets are available from health food shops and can be crushed to feed larvae. Pupae do not feed and should be transferred to clean water to aid their emergence.

As pupae develop the container should either be covered with fine netting held in place with a rubber band or placed inside a mesh cage, so that the emerging adults do not escape. Cages are easily constructed by making a wire frame with a side of about 30 cm and covering this with curtain netting. A circular opening is left in one side, about 10 cm in diameter, and a sleeve is sewn into this to allow access without the mosquitoes escaping. When access is not required the sleeve is knotted.

Adults are fed on sugar solutions soaked into cotton wool. Mosquitoes are very susceptible to desiccation and must be maintained at a high humidity. This can be achieved by covering the cage with a cloth with an end immersed in a trough of water so acting as a wick and enclosing the whole cage in a polythene bag.

Once the adult stage has been reached a decision has to be made as to whether to attempt to obtain eggs from them. Few species will mate readily in confined spaces and there are problems in blood feeding. Some investigators feed the females of man-biting species on themselves but this is a personal decision and there are the obvious risks associated with insect bites. It is possible to feed some species on animals, but this area of investigation is covered by the Animals (Scientific Procedures) Act 1986 and all persons conducting such work must be holders of personal and project licences and undertake the work in premises which have been registered under the Act.

Should you overcome the difficulties of feeding the mosquitoes you still have the problem that they must mate. Fortunately, this can be overcome using an artificial mating technique. This is simple to carry out with the minimum of equipment and produces excellent results. To begin, several males, at least 48 hours old, are anaesthetised with ethyl acetate vapour. They are then fixed with a fast-drying glue to the rim of a Petri dish or a similar container with their heads pointing to the centre of the dish and their undersides uppermost (fig. 37). Using fine scissors, the males are decapitated and the dish is placed beneath a dissecting microscope and viewed at about x10 magnification. A female of the same species is then held by its thorax with a capillary tube connected to a small suction pump. It is moved towards the male so that the tips of the abdomens touch at an angle of about 45° (fig. 37). If this process is observed with a microscope then successful coupling can be witnessed. After separation, the female is transferred to a suitable cage or container.

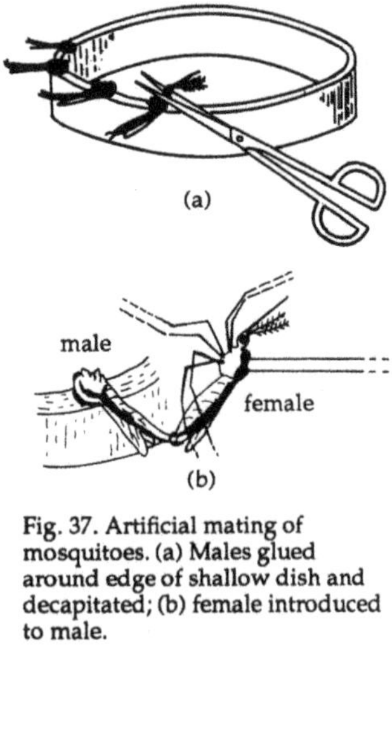

Fig. 37. Artificial mating of mosquitoes. (a) Males glued around edge of shallow dish and decapitated; (b) female introduced to male.

In order to be certain that insemination has been achieved it is advisable to dissect a small number of females and examine their spermatheca(e) for spermatozoa. Details of how to do this are given on p. 59.

Ecological studies

The value of direct observation cannot be overemphasised. We have little knowledge of such things as mating times and circumstances, swarming habits, or overwintering methods including survival of larvae under ice. Also requiring further study are flower preferences for males and females, seasonal emergence patterns of adults and sex ratios of adults at emergence (see Packer & Corbet, 1989b; Snow, 1989). Indeed, many basic aspects of the life cycle of British mosquitoes would benefit from further observations, among them the characteristics and range of larval habitats of given species and instar durations. It is safe to say that any detailed study of an aspect of the biology of mosquitoes of this country will reveal new and interesting information, and may ultimately contribute to the biological knowledge required for the control of disease-carrying mosquitoes in the tropics.

A number of studies can be made by observing mosquito larvae and pupae away from their habitats in tanks or aquaria. As examples, the way in which environmental conditions such as temperature, pH and light regimes affect the development and behaviour of larvae and pupae can be investigated. Also the effects of food availability and overcrowding of larvae on survival and the size of emergent adults can be judged. The usual measurement of adults taken to assess size for this purpose is wing length (see Packer & Corbet, 1989a). Predators such as dytiscid beetle larvae and dragonfly nymphs are believed to have a marked effect on larval and pupal numbers and experiments can be devised to observe and assess this.

By sampling and collecting at regular intervals and noting the instars as well as the species, profiles of the development of the species can also be made. If the water temperature is measured daily then a correlation between this and development periods may also be possible. By comparing such information throughout the country a wider picture of the life cycles of common species can ultimately be achieved.

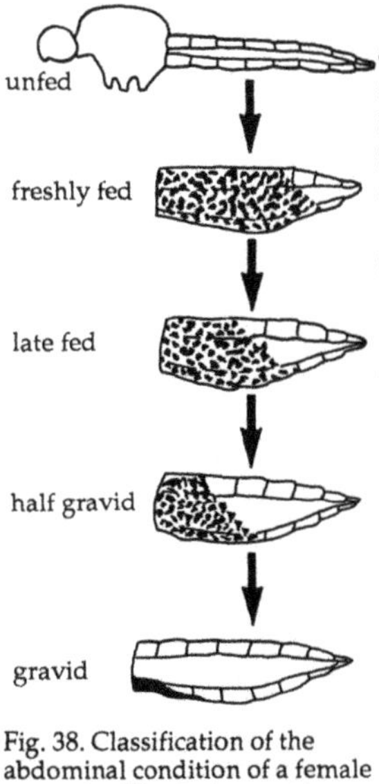

Fig. 38. Classification of the abdominal condition of a female mosquito. The dark area (the blood meal in the stomach) decreases in size as digestion proceeds while the pale area (the developing ovaries) increases in size.

Equally rewarding are observations on adults such as those detailing the timing and duration of the adult season and correlating this with meteorological details; measuring the dispersal from larval habitats; and observing resting sites throughout a given area. All are studies which can be made with the minimum of equipment.

When collecting blood-fed females from resting sites in the field it is interesting to note the phase of digestion and extent of ovarian development. Such observations apply, not only to females collected in the summer and autumn, but also to overwintering mosquitoes. By making such studies the duration of digestion and egg development can be found and compared in different species or in separate localities.

An easy way to record digestion and egg development is to classify the appearance of the abdomen as viewed with a hand lens or dissecting microscope. As this does not entail dissecting the insect, it is obviously the method to use when the insect is to be studied further and allowed to lay its eggs.

The following categories (fig. 38) are used to report the appearance of the abdomen:

unfed – the midgut (stomach) is empty and the abdomen is therefore thin. The ovaries are small, occupying less than one-third of the abdomen. Newly emerged females which have not fed and females which have passed through a cycle of feeding and egg-laying and are awaiting another blood meal are in this category.

freshly-fed – the midgut is distended with red blood and the ovaries occupy about a quarter of the abdomen.

late fed/ half gravid – the blood in the midgut is dark red and the ovaries occupy about half of the abdomen.

gravid – at most a trace of dark blood is visible. The ovaries occupy almost all of the abdomen.

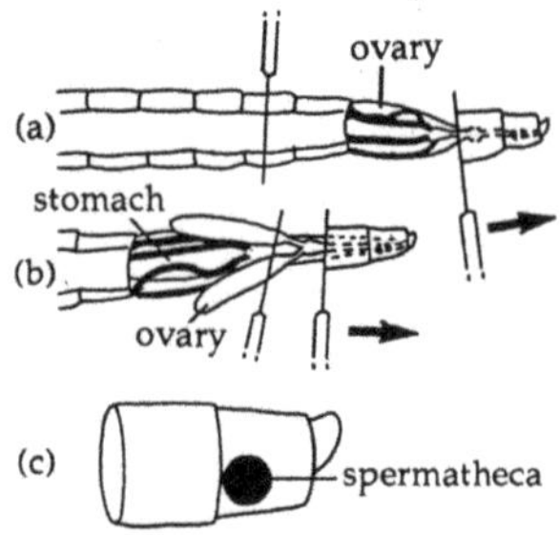

Fig. 39. Removal of the ovaries and spermatheca(e). (a, b) the cuticle is cut between the sixth and seventh abdominal segments and the ovaries extracted; (c) enlarged view of tip of abdomen to show the single spermatheca of *Anopheles*.

For a more detailed look into the extent of ovarian development the mosquito must be dissected to examine the ovaries. This is a relatively simple procedure but one which requires practice and patience. A freshly killed female mosquito is placed on its back in a drop of 0.65% saline solution on a glass slide. With one fine mounted needle on the thorax, another is used to cut the cuticle of the abdomen between the sixth and seventh abdominal segments (fig. 39). The second needle is then used to pull out the ovaries (fig. 39). Once these have emerged, the gut is cut, the ovarian sheath is removed, and the individual strands of the ovary (ovarioles) separated with a needle (fig. 22, p. 14). A cover-slip is placed on the slide and the ovarioles viewed at between x100 and x400 magnification. In addition to the ovaries, the sperm receptacles or spermathecae may also be removed and examined to see whether or not they contain spermatozoa (see below).

The ovaries can then be classified according to a scheme proposed by Christophers (1911). This classification is based upon the development of the most advanced follicles in the ovary (fig. 40). There are five stages:

parous: describes a female which has laid an egg batch. The female may also be referred to as a 'par'

nulliparous: describes a female which has not laid an egg batch. The female may also be referred to as a 'nullipar'

Stage I

The ovarian follicles appear spherical and there are no yolk granules. The developing ovum is small and cannot be distinguished from the accompanying nurse cells (fig. 40). This stage is found in newly emerged females only.

Stage II

The ovarian follicles appear ovoid and yolk can be seen within them to a varying extent. At this stage the ovum occupies up to half of the follicle and its nucleus is visible (fig. 40). Stage II follicles are found in nullipars and in unfed or recently-fed parous females.

Stage III

The yolk occupies about three-quarters of the follicles and the nurse cells the remaining quarter. The nucleus of the ovum is no longer visible (fig. 40). Late-fed and partially gravid females show this ovarian stage.

Stage IV

The ovarian follicles are sausage-shaped and are filled with yolk (fig. 40). This stage is found in females which are becoming gravid.

Stage V

The ova or eggs are fully formed within the follicles. The female is gravid and about to lay her eggs.

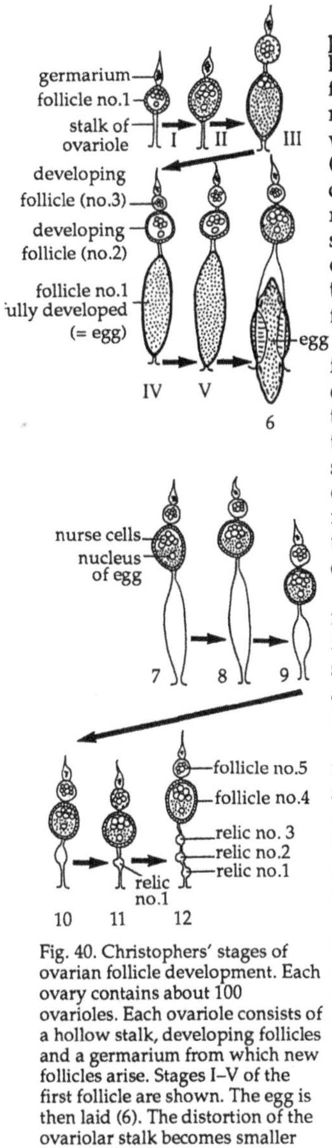

Fig. 40. Christophers' stages of ovarian follicle development. Each ovary contains about 100 ovarioles. Each ovariole consists of a hollow stalk, developing follicles and a germarium from which new follicles arise. Stages I–V of the first follicle are shown. The egg is then laid (6). The distortion of the ovariolar stalk becomes smaller (7–11) but leaves a small dilatation or 'relic'. As successive eggs develop they leave relics from which the number of egg-laying cycles can be determined (12).

As well as deciding upon the ovarian stage, it is also possible to examine the ovary for evidence of past egg-laying cycles. For this examination individual ovarioles from various parts of the ovary are examined at x400 magnification. It is necessary to straighten out the ovarioles with a fine needle to count the number of swellings (dilatations) or to see the presence of a sac. When an egg develops, the ovariole stalk is stretched and this distortion remains, forming a sac, after the egg is laid. Gradually this sac shrinks to leave a small dilatation. By examining an ovariole stalk and counting the number of such dilatations, the number of egg-batches that the mosquito has laid can be found (fig. 40).

Further evidence for deciding whether a female has fed or not can be obtained by examining the tracheal tubes on the surface of the ovary or midgut. The fine tracheal tubes (skeins) are coiled in unfed females but uncoiled when they have fed (fig. 41). Observations on the ovarian tracheal skeins can be made at the same time as the ovaries are examined in the procedure above. The midgut is easy to remove and examine. Following removal of the ovaries and the cutting of the hindgut, the head is removed and the gut can be pulled out from the rear of the body.

In a study of the woodland mosquito *Aedes punctor* in Epping Forest, Essex in 1981, 28% of a population were found to have failed to find a blood meal by the end of the season. Of the remainder, 66% had laid a single egg batch and 12% two egg batches as judged by the above criteria (S.P. Fallis, unpublished).

It is possible to discover whether or not a female has mated. When a female mosquito is inseminated, the spermatozoa are stored in her body and only later are the eggs fertilised. In *Anopheles* there is a single storage sac (spermatheca) in the last segment of the abdomen. In all other genera there are three spermathecae. These sacs are easily recognised by their dark colour (fig. 39).

After removal, the spermathecae are transferred to a drop of water or dilute saline and crushed beneath a cover-slip. If spermatozoa are present they will be forced out and can be recognised by their thread-like appearance and movement.

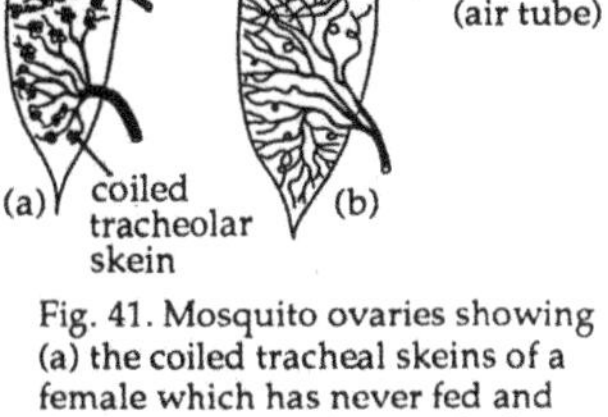

Fig. 41. Mosquito ovaries showing (a) the coiled tracheal skeins of a female which has never fed and (b) the permanently stretched tracheoles of a female which has fed

Table 4. *Observations on swarming and insemination in* Aedes cantans

Year	Mating swarm First observed	Mating swarm Last observed	Date by which more than 90% females inseminated
1980	6 June	23 July	21 May
1981	26 May	29 July	13 May
1982	22 May	2 August	5 May

As the insemination times of most British species are unknown it would be interesting to compile information on this aspect of reproductive biology.

One study by Fallis & Snow (1984) found that nearly all females of *Aedes cantans* were inseminated by the time that the first mating swarms were seen (Table 4).

This raises the question of the precise function of the swarm if, as these results suggest, most females mate before swarms occur. In this species the swarms consist, typically, of 10–40 mosquitoes and form some 75–125 cm above the ground. In late June 1981 the numbers of males in ten swarms were estimated over an evening and the results are shown in fig. 17 (p. 12).

Cytological studies

Mosquitoes are excellent animals for cytological studies. The diploid number of chromosomes is six and salivary gland polytene chromosomes from larvae are large and well banded. In *Anopheles*, aberrations of chromosomes such as inversions, which are typical of certain populations, are easily seen. Few representatives of other genera have been well studied, since their chromosomes are difficult to prepare.

Since the discovery that species of anopheline mosquitoes can be separated by their chromosomal banding patterns, a number of staining techniques have been devised to demonstrate these. In Britain the two members of the *Anopheles maculipennis* complex of species, namely *An. atroparvus* and *An. messeae*, are indistinguishable morphologically, except for egg patterning, but can be recognised by chromosomal banding (White, 1978).

Large active final-instar larvae are chosen and dissected to remove the salivary glands. This is done by first placing the specimen in 5% Carnoy's fixative on a glass slide. The abdomen is cut away with scissor-like movements of a pair of fine needles and the thorax is slit open along the mid-line of the back. By gently pulling on the head, the glands are pulled free of the thorax but are still attached to the head by the salivary ducts. The glands are then cut free and processed as follows:

1 Transfer salivary glands on the tip of a needle to a small drop of Carnoy's fixative on a coverslip for 30–60 seconds.

2 Add about twice the volume of diluted stain (see below) and leave for 3–5 minutes.
3 Place a slide on top of the preparation and tap the coverslip with a finger a few times to break the cell and nuclear membranes and so spread the chromosomes.
4 Wrap the slide and coverslip in tissue paper and press firmly to complete the display of the chromosomes.
5 View at x1000 with an oil immersion lens. Phase contrast is desirable.

The recipes used above are: Carnoy's fixative – 2 parts ethyl alcohol and 1 part glacial acetic acid; concentrated stain – 2% by weight of synthetic orcein powder dissolved in 1 part 85% lactic acid and 1 part glacial acetic acid; dilute stain – 1 part concentrated stain with 1–3 parts 45% acetic acid.

Morphological studies

Many aspects of the morphology of mosquitoes still require further study. Although the basic features are often well documented, all too little is known of character variation within species. This means that many of the ranges for a detail such as the number of pecten spines or the number of comb scales are based on small samples. In *Aedes punctor* larvae the variation in number of pecten spines and comb scales has been investigated and the results are shown in Table 1 (p. 5). When these are being compared for purposes of identification, problems may occur if the ranges quoted are not sufficiently broad. Other features such as variation in adult coloration, scale patterning, size, number, position and form of body hairs also warrant study and require only a basic microscope and slide-making equipment together with time and a desire to investigate.

Where to publish your results

If the results of a study, however small and seemingly insignificant, appear interesting then you may wish to publish them. Several journals exist which publish articles exclusively on insects such as *Entomologist's Monthly Magazine, Entomologist's Gazette, The Entomologist* and *Ecological Entomology*. The newly launched *Dipterists' Digest* caters for a range of articles on all aspects of the biology, ecology and distribution of flies. It is therefore of special relevance to those with an interest in mosquitoes. Depending on the topic and importance of your article you may wish to send it to such a journal for consideration, written, of course, in the house style of the journal. You can judge whether your item will be suitable for a particular journal by comparing your article with those in current issues. It may be that your article is of more local interest in which case it may be valued by a local natural history society for inclusion in its bulletin or newsletter. Some such societies like The London Natural History Society produce a high quality journal, in this case *The London Naturalist*, with contributions of a high standard. Finally there is the British Mosquito Group Newsletter which publishes short notes as well as longer articles on all aspects of the study of British mosquitoes. The address of the British Mosquito Group, which is an outlet for minor and major observations, local records, letters and comments, is given on p. 63.

It is an unbreakable convention of scientific publication that results are reported with scrupulous honesty. Hence it is essential to keep detailed and accurate records throughout the investigation, and to distinguish in the write-up between certainty and probability, and deduction and speculation. In many cases it will be necessary to apply appropriate statistical techniques to test the significance of the findings. A book such as Parker's (1979) *Introductory Statistics for Biology* or the *OU Project Guide* (Chalmers & Parker, 1986) will help, but this is an area where expert advice can contribute much to the planning, as well as to the analysis, of the work.

Useful addresses

Societies and special-interest groups

British Mosquito Group, c/o Dr. N. Burgess, Department of Military Entomology, Royal Army Medical College, Millbank, London, SW1P 4RJ
(a small group of both amateurs and professionals interested in British mosquitoes)

Freshwater Biological Association, The Ferry House, Ambleside, Cumbria, LA22 0LP
(The FBA is the principal Natural Environment Research Council institute studying the biology of freshwaters)

Diptera Recording Scheme, c/o Dr A.E. Stubbs, Nature Conservancy Council, Northminster House, Peterborough, OE1 1UA
(an active scheme supported by a large number of entomologists, both amateur and professional, interested in the ecology and distribution of all kinds of flies)

Royal Entomological Society of London, 41 Queen's Gate, London, SW7 5HU
(the professional organisation for entomologists with an excellent library open to members)

Libraries and booksellers

Natural History Museum Library, Cromwell Road, London, SW7 5BD
(an exceptional library with books on all aspects of natural history)

E.W. Classey Limited, P.O. Box 93, Faringdon, Oxon, SN7 7DR
(sells new and secondhand specialist books on natural history and entomology, often difficult to purchase elsewhere)

Suppliers of equipment

Marris House Nets, 54 Richmond Park Avenue, Bournemouth, BH8 9DR
(suppliers of malaise traps, insect nets and cages)

Watkins and Doncaster, P.O. Box 5, Cranbrook, Kent, TN18 5EZ
(a source of pins, pinning blocks, nets and other useful items for the study of mosquitoes and other insects)

References and further reading

Finding books

Although some of the books and articles listed below and cited in the text will be available in local and school libraries, the majority will not. Specialist libraries, some of which are given in the following section, will be able to assist and your local public library may be able to arrange for you to borrow items from the British Library, Document Supply Centre. This may take several weeks, and it is important to present your librarian with a reference that is correct in every detail. References are acceptable in the form given here, namely the author's name and date of publication, followed by (for a book) the title and publisher or (for a journal article) the title of the article, the full journal title, the volume number, and the first and last pages of the article.

References

Bates, M. (1949). *The Natural History of Mosquitoes*. New York: Harper and Row.

Chalmers, N. & Parker, P. (1986). *OU Project Guide: Fieldwork and Statistics for Ecological Projects*. London: Open University/Field Studies Council.

Chinery, M. (1986). *Collins Guide to the Insects of Britain and Western Europe*. London: Collins.

Christophers, S.R. (1911). The development of the egg follicle in anophelines. *Paludism* **2**, 73–88.

Clements, A.N. (1963). *The Physiology of Mosquitoes*. Oxford: Pergamon.

Cranston, P.S., Ramsdale, C.D., Snow, K.R. & White, G.B. (1987). *Keys to the British Mosquitoes* (Culicidae). Freshwater Biological Association Scientific Publications, **48**.

Curtis, C.F. (1986). Fact and fiction in mosquito attraction and repulsion. *Parasitology Today* **2**, 316–318.

Davis, E.E. (1985). Insect repellents: concepts of their mode of action relative to potential sensory mechanisms in mosquitoes (Diptera: Culicidae). *Journal of Medical Entomology* **22**, 237–243.

Disney, R.H.L. (1970). A note on the uptake of sugar by some blood-sucking flies. *Proceedings of the Royal Entomological Society of London (A)* **45**, 51–54.

Fallis, S.P. & Snow, K.R. (1983a). The hatching stimulus for eggs of *Aedes punctor* (Diptera: Culicidae). *Ecological Entomology* **8**, 23–28.

Fallis, S.P. & Snow, K.R. (1983b). Distribution of the eggs of *Aedes punctor*. *Ecological Entomology* **8**, 139–144.

Fallis, S.P. & Snow, K.R. (1983c). Larval development in *Orthopodomyia pulchripalpis* (Rondani) (Diptera: Culicidae). *Entomologist's Monthly Magazine* **119**, 75–79.

Fallis, S.P. & Snow, K.R. (1984). Observations on the swarming of *Aedes cantans* Meigen (Diptera: Culicidae). *Entomologist's Gazette* **35**, 57–60.

Fish, D. & Joslyn, D.J. (1984). Larval population estimates of *Aedes communis* using Giemsa marking. *Mosquito News* **44**, 565–567.

Foster, W.A. & Lutes, K.I. (1985). Tests of ultrasonic emissions on mosquito attraction to hosts in a flight chamber. *Journal of the American Mosquito Control Association* **1**, 199–202.

Gillett, J.D. (1971). *Mosquitos*. London: Weidenfeld and Nicolson.

Gomulski, L. (1988). Observations on the swarming behaviour of three species of *Aedes* mosquitoes in Britain. *British Mosquito Group Newsletter* **5**, 5–7.

Guthrie, M. (1989). *Animals of the Surface Film*. Naturalists' Handbooks 12. Slough: Richmond Publishing Co. Ltd.

Harbach, R.E. & Knight, K.L. (1980). *Taxonomists' Glossary of Mosquito Anatomy*. New Jersey: Plexus.

Joslyn, D.J., Conrad, L.B. & Slavin, P.T. (1985). Development and preliminary field testing of the Giemsa self-marker for the salt marsh mosquito *Aedes sollicitans* (Walker) (Diptera: Culicidae). *Annals of the Entomological Society of America* **78**, 20–23.

Kitching, R.L. (1971). An ecological study of water-filled tree holes and their position in the woodland ecosystem. *Journal of Animal Ecology* **40**, 281–302.

Knight, K.L. & Stone, A. (1977). *A Catalog of the Mosquitoes of the World* (Diptera: Culicidae). Thomas Say Foundation.

Marshall, J.F. (1938). *The British Mosquitoes*. London: British Museum (Natural History). (Johnson reprint available).

Mattingly, P.F. (1950). Family Culicidae. Subfamily Culicinae. In *Diptera. 2. Nematocera. Handbooks for the Identification of British Insects* **9** (2), pp. 102–120. Royal Entomological Society of London.

Mattingly, P.F. (1969). *The Biology of Mosquito-borne Disease*. London: Allen and Unwin.

Packer, M.J. & Corbet, P.S. (1989a). Size variation and reproductive success of female *Aedes punctor* (Diptera: Culicidae). *Ecological Entomology* **14**, 297–309.

Packer, M.J. & Corbet, P.S. (1989b). Seasonal emergence, host-seeking activity, age composition and reproductive biology of the mosquito *Aedes punctor. Ecological Entomology* **14**, 433–442.

Parker, R.E. (1979). *Introductory Statistics for Biology* (Studies in Biology). London: Edward Arnold.

Ramsdale, C.D. & Smith, G. (1990). A record of *Aedes vexans* from Kent. *British Mosquito Group Newsletter* **7**, 2.

Ramsdale, C.D. & Wilkes, A.J. (1985). Some aspects of overwintering in southern England of the mosquitoes *Anopheles atroparvus* van Thiel, 1927 and *Culiseta annulata* Schrank, 1776 (Diptera: Culicidae). *Ecological Entomology* **10**, 449–454.

Rees, A.T. & Rees, A.E. (1989). *Anopheles algeriensis* on Anglesey: the story so far. *British Mosquito Group Newsletter* **6**, 1–5.

Schreck, C.E. (1985). The status of Deet (N, N-Diethyl-m-toluamide) as a repellent for *Anopheles albimanus. Journal of the American Mosquito Control Association* **1**, 98–100.

Service, M.W. (1968a). Observations on feeding and oviposition in some British mosquitoes. *Entomologia Experimentalis et Applicata* **11**, 277–285.

Service, M.W. (1968b). The taxonomy and biology of two sympatric sibling species of *Culex, C. pipiens* and *C. torrentium* (Diptera: Culicidae). *Journal of Zoology* **156**, 313–323.

Service, M.W. (1969). Observations on the ecology of some British mosquitoes. *Bulletin of Entomological Research* **59**, 161–194.

Service, M.W. (1971a). Feeding behaviour and host preferences of British mosquitoes. *Bulletin of Entomological Research* **60**, 653–661.

Service, M.W. (1971b). The daytime distribution of mosquitoes resting amongst vegetation. *Journal of Medical Entomology* **8**, 271–278.

Service, M.W. (1973). The biology of *Anopheles claviger* Meigen (Diptera: Culicidae) in southern England. *Bulletin of Entomological Research* **63**, 347–359.

Service, M.W. (1976). *Mosquito Ecology. Field Sampling Methods*. London: Applied Science Publishers.

Service, M.W. (1977). Ecological and biological studies on *Aedes cantans* (Meigen) (Diptera: Culicidae) in southern England. *Journal of Applied Ecology* **14**, 159–186.

Service, M.W. (1984). Evaluation of sticky light traps for sampling mosquito larvae. *Entomologia Experimentalis et Applicata* **35**, 27–32.

Snow, K.R. (1984a). The identification of larvae of British mosquitoes (Diptera: Culicidae). *Entomologist's Monthly Magazine* **120**, 21–32.

Snow, K.R. (1984b). Evaluation of *Bacillus thuringiensis*, serotype H-14 formulations as larvicides for *Aedes punctor* (Kirby). *International Pest Control* **26**, 12–14.

Snow, K.R. (1986). Mosquito nuisance and control in U.K. *Environmental Health* **94**, 265–268.

Snow, K.R. (1987a). Towards a check-list of British mosquitoes. *Entomologist's Monthly Magazine* **123**, 83–89.

Snow, K.R. (1987b). The names of British mosquitoes. *Entomologist's Monthly Magazine* **123**, 127–139.

Snow, K.R. (1987c). Control of mosquito nuisance in Britain. *Journal of the American Mosquito Control Association* **3**, 271–275.

Snow, K.R. (1987d). Emergence patterns of *Aedes punctor* and *Aedes cantans* (Diptera: Culicidae) in southern England. *Entomologist's Gazette* **38**, 253–262.

Snow, K.R. (1989). The use of pupal exuviae in the study of seasonal emergence patterns and sex ratios. *British Mosquito Group Newsletter* **6**, 6.

White, G.B. (1978). Systematic reappraisal of the *Anopheles maculipennis* complex. *Mosquito Systematics* **10**, 13–44.

Yates, M.G. (1979). Biology of the treehole breeding mosquito *Aedes geniculatus* in southern England. *Bulletin of Entomological Research* **69**, 611–628.

Young, A.D.M. & Downe, A.E.R. (1982). Renewal of sexual receptivity in mated female mosquitoes, *Aedes aegypti. Physiological Entomology* **7**, 467–471.

Index

CPSIA information can be obtained
at www.ICGtesting.com
Printed in the USA
BVHW092112161219
566718BV00003B/63/P

9 781784 270476